Leckie
the education publisher
for Scotland

C000231320

Higher
HUMAN BIOLOGY
For SQA 2019 and beyond

Student Book

Billy Dickson, Graham Moffat

001/100820

10 9 8 7 6 5 4 3 2 1

ISBN 9780008384449

Published by
Leckie
An imprint of HarperCollins Publishers
Westerhill Road, Bishopbriggs, Glasgow, G64 2QT
T: 0844 576 8126   F: 0844 576 8131
leckiescotland@harpercollins.co.uk   www.leckiescotland.co.uk

Publisher: Sarah Mitchell
Project manager: Kerry Ferguson and Lauren Murray

Special thanks to
Jouve (layout and illustration)
Jess White (proofread)

Printed in Italy by GRAFICA VENETA S.p.A.

A CIP Catalogue record for this book is available from the British Library.

Acknowledgements
Whilst every effort has been made to trace the copyright holders, in cases where this has been unsuccessful, or if any have inadvertently been overlooked, the Publishers would gladly receive any information enabling them to rectify any error or omission at the first opportunity.

Leckie would like to thank the following copyright holders for permission to reproduce their material:

Figure 1.1.7a *Tewan Banditrukkanka / Shutterstock*; Figure 1.1.7b *Mauro Fermariello / Science Photo Library*; Figure 1.1.7c *Southern Illinois University / Science Photo Library*; Figure 1.2.1a *Barrington Brown / Science Photo Library*; Figure 1.2.1b *Donaldson Collection / Contributor*; Figure 1.8.3a *Sergii Rudiuk / Shutterstock.com*; Figure 1.8.4a *Maxisport / Shutterstock.com*; Figure 3.3.4a *Marek Valovic / Shutterstock.com*; Figure 3.5.2 *Blamb / Shutterstock*; Figure 3.5.4a *Scott Camazine / Alamy*; P282 © *Scottish Qualifications Authority*

All other images © Shutterstock.com

Chapter 3

Neurobiology and immunology

Welcome to your Higher Human Biology Student Book!

This book covers all of the knowledge, understanding and skills of scientific inquiry included in the SQA Higher Human Biology course. It has been designed to help you pass and achieve your best grade in the course.

It is recommended that you download and save a copy of the course specification from the SQA website at: www.sqa.org.uk/sqa/47915.html. There is other useful information on this page, including past papers and their marking instructions and information relating to coursework, specifically the assignment.

Course assessment

There are three components in the assessment of the SQA Higher Human Biology course:

- **Paper 1:** An objective question paper to be completed in 40 minutes. It is a multiple-choice question paper with 25 items worth 1 mark each. Total = 25 marks.
- **Paper 2:** A structured and extended response question paper to be completed in 2 hours and 20 minutes. It is a written question paper, containing about 14–16 questions of structured and extended responses. Total = 95 marks.
- **Assignment report:** A report to be written in school under controlled conditions in a maximum of 2 hours. It is marked out of 20 marks by SQA and will be scaled up to 30 marks so that it makes up 20% of your overall course assessment and grading.

The total number of marks available for the course is 150 marks.

- You must achieve 50% of the total marks for an award at grade C (75 out of 150 marks).
- You must achieve 60% of the total marks for an award at grade B (90 out of 150 marks).
- You must achieve 70% of the total marks for an award at grade A (105 out of 150 marks).

Features of this book

YOU SHOULD ALREADY KNOW

You should already know:

- The meanings of the terms 'haploid' and 'diploid'.
- Most cells are diploid but gametes are haploid.

Each chapter starts with a summary of the assumed prior knowledge and skills gained from study at National 5 level.

LEARNING INTENTIONS

Learning intentions

- Describe cell division in somatic and germline cells.
- Explain the process of cellular differentiation.

This section contains a broad list of the content of the key area covered by the chapter.

MAKE THE LINK

This feature identifies links to related content within and across key areas in Higher Human Biology. These links are worth following because many questions in your exam are integrated within and across different areas.

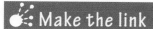

Make the link

There is more about acetylcholine in Chapter 2.6 on page 156.

KEY WORDS

Key words are in **bold** text. You must know the meaning of these words to do well in Higher Human Biology. **Red bold** terms have a glossary box with a definition of the term. **Black bold** terms are also key terms but their definition is made clear in the text.

Cancer cell

A cell that grows and divides in an unregulated way to produce a tumour.

HINT

These boxes contain tips to support learning or related to exam techniques.

Hint

Antiparallel is a tricky idea – remember the strands are like the two lanes of traffic travelling different ways on a road: the same but going in the opposite direction.

TECHNIQUES LOGO

This symbol indicates a technique with which you are expected to be familiar for your exam.

Gel electrophoresis

Once DNA has been amplified by PCR it can be treated with enzymes to produce specific fragments and then the fragments can be stained ready for gel electrophoresis.

ASSIGNMENT SUPPORT LOGO

This symbol indicates a procedure or technique which could be used to generate data for an assignment.

Assignment Support

ACTIVITIES

There are sets of activities designed to support learning. They are separated into those done individually, in pairs or in groups. The individual activities are designed to help reinforce learning and to evaluate progress. You are encouraged to mark you own work using the answers online at www.collins.co.uk/pages/Scottish-curriculum-free-resources. Your responses should help you to evaluate your overall grasp of the key area. Paired and group activities are designed to reflect success criteria and to support learning, as well as to tap into the value of learning cooperatively. All online activities can be found at www.collins.co.uk/pages/Scottish-curriculum-free-resources.

Work individually to...

GO! Activity 1.2.1 Work individually to ...

Structured questions

1. DNA is the substance which carries the human genetic code.

   a) **Describe** the shape of a DNA molecule.   1

   b) **Explain** the role of hydrogen bonds in DNA structure.   1

   c) **Name** the components of a single nucleotide.   1

Structured questions: These are short - answer questions to test your knowledge and understanding of the key area. The final questions in these sections test scientific inquiry skills such as planning or designing experiments to test given hypotheses, selecting information from a variety of sources, presenting information appropriately in a variety of forms, processing information using calculations and units where appropriate, making predictions and generalisations based on evidence, drawing valid conclusions and giving explanations supported by evidence, and evaluating by suggesting improvements to experiments and investigations. All skills are covered repeatedly over the whole book.

GO! Activity 1.2.1 Work individually to ...

Extended response questions

1. **Give** an account of stem cells under the following headings:

   a) Properties of embryonic stem cells   4

   b) Stem cells in adult blood tissue   4

Extended response questions: There is a set of extended response questions for between 4 and 10 marks testing related knowledge in the key area. There is usually one mini extended response question for about 4–5 marks and a full extended response question for between 6 and 10 marks.

It is recommended that, as a general rule, you spend no more than 90 seconds on each available mark. For example, if a question is worth 4 marks, you should spend no more than 6 minutes in producing your answer.

Work in pairs to...

These active learning tasks allow you to work with a partner and include appropriate **experimental and observational work** and activities, such as **card sorting**, **card sequencing** and **formative assessment opportunities** such as **Dice and Slice**. There is always a **flashcard activity** to be done!

> **GO!** Activity 1.1.2 Work in pairs to...
>
> **1. Practical activity: Examining meiotic cells from testes**
>
> *You will need: a microscope, a prepared slide of testes tissue (this may not be from a human source but the basic structure is similar) and you will both need a sharp pencil, a rubber and a piece of A5 blank paper.*

Work as a group to...

These activities promote cooperative learning in both small and larger groups. They are designed to encourage discussion of the learning intentions and the use of appropriate vocabulary for success. There are some **online research tasks**, usually with a presentation to prepare, poster or model construction as well as some consolidation challenges: **Ring of Fire** is a whole class activity designed to check and consolidate the knowledge and understanding content covered in the key area; **Placemat** is a cooperative learning strategy which allows you to think about, record and share your ideas in groups.

> **GO!** Activity 1.2.3 Work as a group to ...
>
> **1. Design and make a model of DNA.**
>
> *You will need: two different colours of miniature marshmallow sweets to represent sugars and phosphates, four colours of jelly sweets for the bases and cocktail sticks cut in two to join the components.*
>
> Arrange five nucleotide pairs like a ladder – don't try to model the double helix!

SUCCESS CRITERIA

Each chapter closes with a concise list of success criteria covering knowledge and skills and detailing what you should be able to do when you have completed the chapter. You should use these lists to help evaluate where you are with your learning and indicate the next steps needed to make any improvements. There is a traffic light tool to help with this and your evaluation should be based on your success with the questions in the individual activities. Don't be afraid to give criteria red lights – that's part of learning! Just make sure you ask your teacher for help with any that you give a red light.

PRACTICE AREA TESTS

At the end of chapters 1–3 there are 50-mark tests. Each test is made up of two papers – Paper 1 with 10 multiple-choice items and Paper 2 with 40 marks of structured and extended response questions. Each paper has a balance of knowledge and understanding and skills questions. We suggest you allow 75 minutes to complete each test. The answers are provided online to help you mark your own work. You could also grade these tests – as a rough guide:

- 25 marks = C
- 30 marks = B
- 35 marks = A

SKILLS OF SCIENTIFIC INQUIRY

Chapter 4 deals with the skills of scientific inquiry, which make up about 25% of the question paper marks for the Higher Human Biology course. Each of the seven named skills are dealt with in turn through descriptions and worked examples. Hints are given throughout the chapter. You should note that, in every key area, the last question or two of the individual activities concentrate on these scientific inquiry skills.

YOUR ASSIGNMENT

Chapter 5 deals with the assignment and its report, which is worth 20% of the overall marks for the Higher Human Biology course. The research and report stages are dealt with in turn and hints are given throughout.

ANSWERS

The answers online are national standard responses to all of the activity questions in each key area as well as solutions to some other activities such as sequencing tasks. It is highly recommended that you use these answers to mark your own work and to consolidate and evaluate your learning.

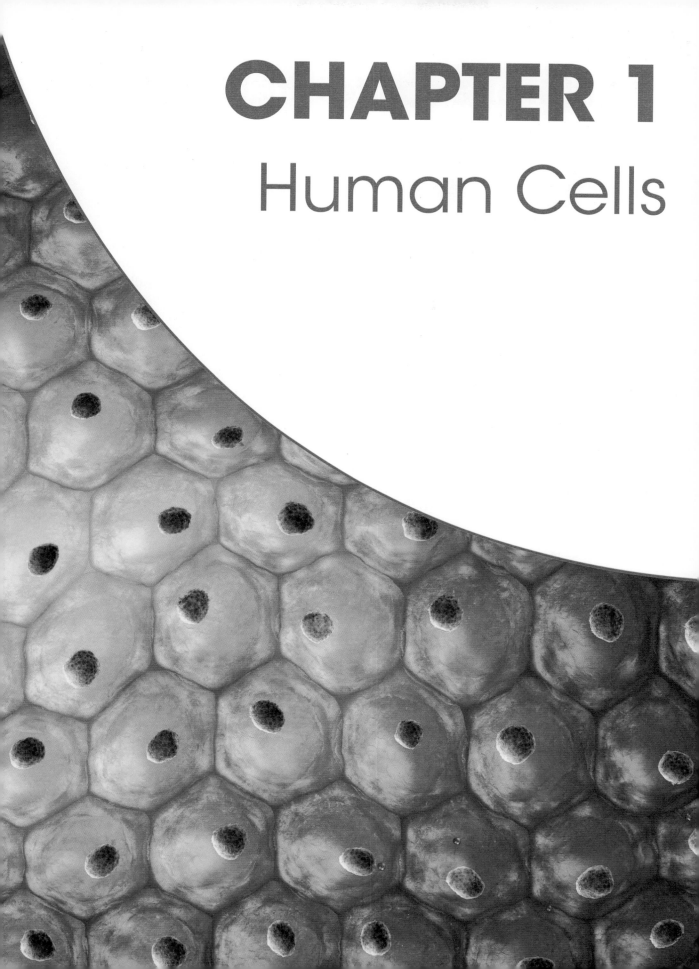

CHAPTER 1
Human Cells

1.1 Division and differentiation in human cells

You should already know:

- The meanings of the terms 'haploid' and 'diploid'.
- Most cells are diploid but gametes are haploid.
- The basic structures of sperm and egg cells.
- Human sperm cells are produced in testes and egg cells are produced in ovaries.
- The nuclei of haploid gametes fuse to produce a diploid zygote during fertilisation.
- The sequence of events of mitosis and the terms 'chromatids', 'equator' and 'spindle fibres'.
- Mitosis provides new cells for growth and repair of damaged cells and maintains the diploid chromosome complement.
- Stem cells in humans are unspecialised cells which can divide in order to self-renew.
- Stem cells have the potential to become different types of cell and are involved in growth and repair.
- Specialisation of cells leads to the formation of a variety of cells, tissues and organs.
- Organs work together to form systems.

Learning intentions

- Describe cell division in somatic and germline cells.
- Explain the process of cellular differentiation.
- Describe the therapeutic and research uses of stem cells.
- Describe cancer cells and explain how they can form tumours and secondary tumours.

📖 Somatic cell

A body cell which can divide by mitosis to form more body cells.

📖 Diploid

Referring to a human body cell which has two matching sets of chromosomes.

📖 Homologous chromosomes

Chromosomes which carry the same sequence of genes and other DNA sequences.

Division of somatic and germline cells

Somatic cells make up the bulk of the human body and are nearly all **diploid**. This means that their nuclei each contain two matching sets of 23 chromosomes, one from each parent, giving a total chromosome complement of 46. **Figure 1.1.1** shows the chromosomes from an adult female somatic cell. Note that each chromosome from one of the sets is matched with one from the other set. A pair of matching chromosomes are said to be **homologous chromosomes**. Note also that each chromosome is replicated – its DNA has been copied and the chromosome appears as a double strand made up of two identical **chromatids**. Somatic cells undergo **mitosis** in the production of new somatic cells during the growth, development and repair of the body.

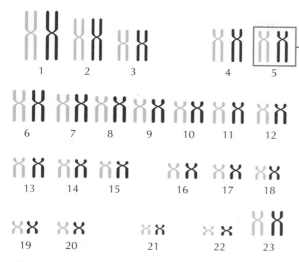

One homologous pair of chromosomes

Figure 1.1.1 *The diploid chromosome complement from a human somatic cell – note that each chromosome in a set is matched with its homologous chromosome from the other set and each chromosome has been replicated into two chromatids*

Germline cells found in human ovaries and testes are involved in reproduction. They undergo mitosis to produce more germline cells, and another type of cell division, called **meiosis**, to produce gametes. Male gametes are sperm and female gametes are ova. **Figure 1.1.2** shows the movements of chromosomes during mitosis and meiosis.

Make the link

There is more about DNA replication in Chapter 1.2 on page 26.

Make the link

There is more about ovaries and testes in Chapter 2.1 on page 100.

Germline cell

A cell which can give rise to gametes.

(a)

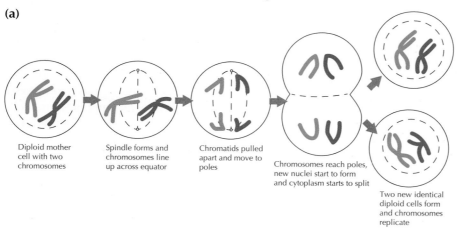

(b)

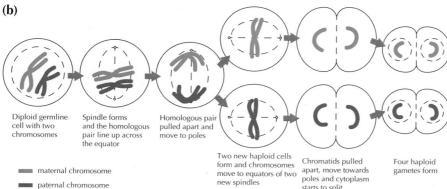

Figure 1.1.2 *(a) Mitosis in a cell with diploid number two - cell divides to produce two identical diploid daughter cells*
(b) Meiosis in a cell with diploid number two - cell divides to produce four haploid gametes

Gametes are **haploid** and contain only one set of 23 chromosomes. When the nuclei of a male and female gamete fuse together during fertilisation, a zygote is formed containing the diploid number of chromosomes. **Figure 1.1.3** shows an outline of the human life cycle.

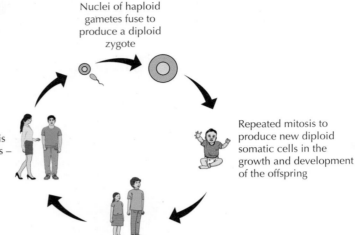

Nuclei of haploid gametes fuse to produce a diploid zygote

Repeated mitosis to produce new diploid somatic cells in the growth and development of the offspring

Diploid germline cells undergo mitosis and meiosis to produce haploid gametes – sperm in males and ova in females

Figure 1.1.3 *Outline of the human life cycle giving the roles of mitosis and meiosis*

Cellular differentiation

The human body, from its time as an embryo and on into adulthood, contains some cells which are unspecialised. These unspecialised cells are called stem cells. All the genes in **embryonic stem cells** can be switched on so these cells can differentiate into any type of cell needed during the growth and development of an embryo. These cells can be described as **pluripotent** and this idea is shown in **Figure 1.1.4**.

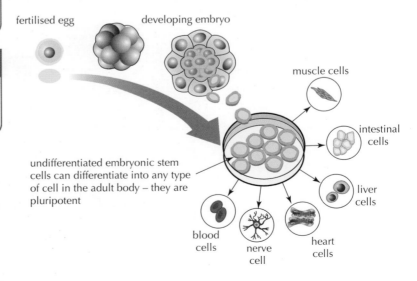

fertilised egg

developing embryo

muscle cells

intestinal cells

liver cells

heart cells

nerve cell

blood cells

undifferentiated embryonic stem cells can differentiate into any type of cell in the adult body – they are pluripotent

Figure 1.1.4 *The potential of embryonic stem cells*

Stem cells in the adult body are called **tissue stem cells**. These are **multipotent** and they can differentiate into all of the types of cell found in a particular tissue type. For example, blood stem cells located in bone marrow can give rise to the different cellular components of blood: red blood cells, platelets, phagocytes and lymphocytes. This idea is shown in **Figure 1.1.5**.

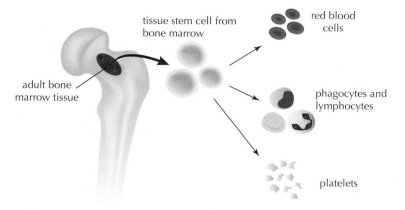

Figure 1.1.5 *The potential of tissue stem cells from an adult – these are in bone marrow and have the potential to become any type of blood cell*

In cellular differentiation, a cell starts to express certain genes to produce the proteins characteristic for that type of cell. It is the presence of these proteins that allows the cell to carry out its specialised functions. This is shown in **Figure 1.1.6**.

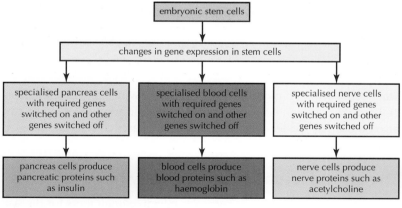

Figure 1.1.6 *Chart outlining the changes in gene expression which occur during cell differentiation*

📖 Tissue stem cell

Multipotent cell from an adult tissue which can differentiate into any type of cell from that tissue.

📖 Multipotent

Refers to a cell which has the potential to differentiate into the range of cells from a particular tissue.

Make the link

There is more about phagocytes and lymphocytes in Chapter 3.5 on page 225.

Make the link

There is more about gene expression in Chapter 1.3 on page 35.

Make the link

There is more about insulin in Chapter 2.8 on page 175.

Make the link

There is more about haemoglobin in Chapter 2.4 on page 132.

Make the link

There is more about acetylcholine in Chapter 2.6 on page 156.

Therapeutic and research uses of stem cells

Therapeutic uses

Therapeutic uses of stem cells are those used in the medical treatment of patients. They involve the repair of damaged or diseased organs or tissues. Treatments which currently use stem cell therapy include corneal repair and the regeneration of damaged skin as shown in **Figure 1.1.7**.

Stem cells from the embryo can self-renew, under the right conditions, *in vitro* laboratory conditions. The pluripotent potential of these cells makes them ideal for use in therapies. However, to obtain embryonic stem cells for *in vitro* self-renewal, the donor embryo is destroyed. The use of embryonic stem cells in this way is highly regulated in the UK and the whole area of this type of work raises serious ethical issues (see below).

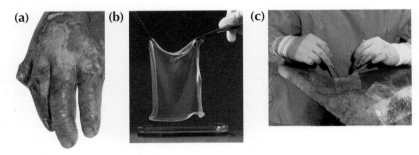

(a)   **(b)**   **(c)**

Figure 1.1.7 *Therapeutic use of stem cells in skin grafting. (a) Patient with a seriously burned hand (b) Skin tissue stem cells cultured* in vitro *to produce a new layer of skin (c) New layer grafted onto patient's affected area*

Research uses

Stem cell research provides information on how cell processes such as cell growth, differentiation and gene regulation work. Stem cells can also be used as model cells to study how diseases develop or be used for drug testing.

The ethical issues of using embryonic stem cells

Use of embryonic stem cells can offer effective treatments for disease and injury, however, it involves destruction of embryos. This can raise an ethical dilemma because in medicine there is a duty to relieve suffering but also to preserve life. At the present time, stem cell research is highly regulated in the UK and stem cell treatments are carried out using tissue stem cells.

The potential therapeutic uses of embryonic stem cells

Stem cells are currently involved in transplantation procedures such as bone marrow transplant and skin and corneal grafts. However, they have the potential to provide a supply of material for transplantation

Figure 1.1.8 *Will doctors be able to offer lifelong stem cell treatments for type 1 diabetes?*

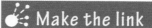

therapies to treat a much wider range of conditions. Some human conditions are caused by the death or failure of certain cell types. Insulin-producing cells in type 1 diabetes and dopamine-sensitive cells in Parkinson's disease are examples. Stem cell transplants have the potential to treat and perhaps even cure these and other conditions.

> ### Make the link
>
> There is more about type 1 diabetes in Chapter 2.8 on page 177.

Cancer

Cells respond to chemical signals which help to keep them functioning normally. **Cancer cells** are those which stop responding to these **regulatory signals** and may start to divide excessively. This results in a mass of abnormal cells called a tumour. Cells within the tumour may fail to attach to each other properly, spreading through the body where they may settle and continue to divide, forming **secondary tumours** as shown in **Figure 1.1.9**.

> ### 📖 Cancer cell
>
> A cell that grows and divides in an unregulated way to produce a tumour.

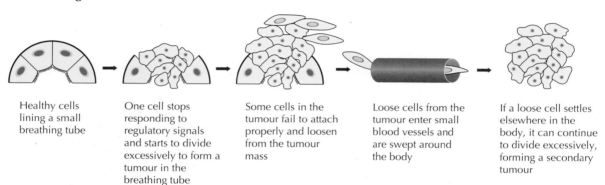

Healthy cells lining a small breathing tube

One cell stops responding to regulatory signals and starts to divide excessively to form a tumour in the breathing tube

Some cells in the tumour fail to attach properly and loosen from the tumour mass

Loose cells from the tumour enter small blood vessels and are swept around the body

If a loose cell settles elsewhere in the body, it can continue to divide excessively, forming a secondary tumour

Figure 1.1.9 *Formation of a tumour and the development of a secondary tumour in cancer*

🔵 Activity 1.1.1 Work individually to …

Structured questions

1. The human body contains both somatic and germline cells.
   a) **Describe** the difference between a somatic cell and a germline cell.   2
   b) **Name** the **two** types of cell division carried out by germline cells.   1
   c) Compare the chromosome complement of a somatic cell with that of a gamete.   1
2. **Describe** what is meant by a homologous pair of chromosomes.   1
3. Re-write the following phrases describing stages in meiosis in the correct order of their occurrence.   1

   Homologous chromosomes separate

   Haploid gametes form

   Sister chromatids separate

   Homologous pairs form
4. a) **State two** characteristics of cancer cells.   2
   b) **Describe** how a secondary tumour can form.   2
5. **Explain** why the use of embryonic stem cells has caused an ethical dilemma.   2

6. Cancer patients can be treated using chemotherapy.

This treatment destroys tumour cells but also reduces the number of white blood cells. As a result, patients have an increased chance of infection.

The graph shows the white blood cell count of a cancer patient and their increased chance of infection in the days following chemotherapy treatment.

 Make the link

There is information about double axis line graphs in Chapter 4 on page 266.

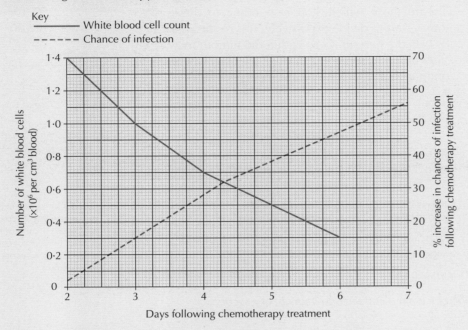

a) **Calculate** the average decrease in number of white blood cells per day between Day 2 and Day 5. 1

b) **State** the number of days following chemotherapy by which the chance of infection increased to 25%. 1

c) **Give** the percentage increase in chance of infection when the number of white blood cells was $1 \cdot 0 \times 10^6$ per cm^3 of blood. 1

d) **Predict** the number of white blood cells which would be expected per cm^3 of blood 7 days after the chemotherapy treatment. 1

Extended response questions

1. **Give an account** of stem cells under the following headings:
   a) Properties of embryonic stem cells 4
   b) Stem cells in bone marrow 4

2. **Write notes** on the uses of human stem cells under the following headings:
   a) Therapeutic uses of stem cells 3
   b) Research uses of stem cells 3

GO! Activity 1.1.2 Work in pairs to …

1. **Practical activity: Examining meiotic cells from testes**

   You will need: a microscope, a prepared slide of testes tissue (this may not be from a human source but the basic structure is similar) and you will each need a sharp pencil, a rubber and a piece of A5 blank paper.

   Method

   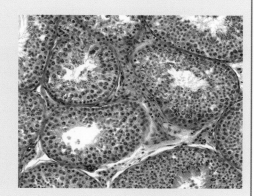

   - Set up your microscope as directed by your teacher.
   - Place the prepared slide of testes tissue onto the stage and focus at high magnification on the preparation. If you can't do this, use the photograph here on the right.
   - Look at the individual cells. The male gamete mother cells are found around the edges of the tubes which are cut through in the preparation. Germline cells are undergoing meiosis nearer to the centre of the tubes and the gametes (sperm cells) are found free in the centre of the tubes.

   a) Using a piece of A5 blank paper and a sharp pencil, draw a gamete mother cell, a germline cell with chromosomes visible and a gamete (sperm cell) and label each one. Glue or tape your diagram into your notes.

   b) Answer the following questions:

      i. **Name** the tissue from which the preparation comes.   1

      ii. This tissue contains germline cells.

         Describe what is meant by the term 'germline'.   1

      iii. **Name** the gametes which are being made here.   1

      iv. The gamete mother cell is diploid and the gametes are haploid.

         Explain the difference between diploid and haploid.   2

2. **Research how the body replaces blood.**

   Visit the NHS web page below:

   www.blood.co.uk/the-donation-process/after-your-donation/how-your-body-replaces-blood/

   Discuss the material on the web page and note down the answers to the following questions:

   a) **Give** the percentage of the body that is made up of blood.

   b) **Describe** how the bone marrow stem cells are triggered to start producing red blood cells instead of white blood cells or platelets.

   c) **Identify** the rate at which new red blood cells can be made.

   d) **Identify** the recommended time needed between blood donations.

   e) **Identify** the organs in which iron is stored to be used in making new haemoglobin for red blood cells.

3. **Flashcard activity**

   You will each need: a set of blank flashcards (A7 cards) and a stopwatch.

   - Find the glossary terms for this chapter – they are the **black** typeface and red typeface terms. Using your blank cards, you should each make a set of flashcards for these terms – write the term on one side and the definition on the other. You will find the definitions in the chapter.

- Shuffle your cards and lay them out in a column, some showing terms and some showing definitions – you decide. Your partner should match their cards with yours, laying their cards in a column beside yours to give the corresponding term or definition. Time how long they take to do this.
- Now swap roles – your partner should lay out their cards and you should try the matching exercise while your partner times you.
- You should each keep your set of flashcards as a revision tool for later.

GO! Activity 1.1.3 Work as a group to ...

1. **Design and make a model of meiosis in a cell with a diploid number of four.**

    You will need: a few blue and a few yellow pipe-cleaners, scissors, glue stick, a piece of A3 card and a marker pen.

    Read through the steps below and divide the tasks between the group:

    - Use the pipe-cleaners to make four small (3–4 cm) chromosomes: one smaller blue, one smaller yellow, one larger blue and one larger yellow – ensure that each is made up of two sister chromatids. You can twist two chromatids together to make a chromosome.
    - On the A3 card, draw out the shapes of cells as shown in Figure 1.1.2(b) on page 13.
    - Make pipe-cleaner models of chromosomes and chromatids for all the remaining stages.
    - Put the model together and glue or tape the chromosomes into position on the card.

2. **Investigate and debate the ethics surrounding stem cell research and the sources of stem cells.**

    You will need: a few pieces of A6 card.

    - Access the website below and view the video clip which is linked to the page. Your teacher may show this clip to the whole class:
      www.eurostemcell.org/embryonic-stem-cell-research-ethical-dilemma
    - Use pieces of card to note down the points you find most persuasive – either for or against the use of embryonic stem cells in medicine.
    - Your teacher may organise a debate on the uses of embryonic stem cells. Be sure you know where you stand on this issue and have some reasons on your cards to support your position.

Learning checklist

After working on this chapter, I can:

Knowledge and understanding

1. State that a somatic cell is any cell in the body other than cells involved in reproduction.

2. State that somatic stem cells divide by mitosis to form more somatic cells.

3. State that germline cells are the stem cells that divide to form gametes.

4. State that germline stem cells divide by mitosis to produce more germline stem cells and by meiosis to produce haploid gametes.

5. Describe a germline stem cell dividing by mitosis to maintain the diploid chromosome number of 23 pairs of homologous chromosomes.

6. Describe a germline stem cell dividing by meiosis by undergoing two divisions, first separating homologous chromosomes and second separating sister chromatids to produce haploid gametes containing 23 single chromosomes.

7. Explain that cellular differentiation is the process by which a cell expresses certain genes to produce proteins characteristic for that type of cell to allow the cell to carry out specialised functions.

8. State that cells in the very early embryo can differentiate into all the cell types that make up the individual – they are pluripotent.

9. Explain that all the genes in embryonic stem cells can be switched on so these cells can differentiate into any type of cell.

10. State that tissue stem cells are involved in the growth, repair and renewal of the cells found in that tissue – they are multipotent.

11. Explain that tissue stem cells are multipotent as they can differentiate into all of the types of cell found in a particular tissue type.

12. Describe blood stem cells located in bone marrow as giving rise to red blood cells, platelets, phagocytes and lymphocytes.

13. State that the therapeutic use of stem cells involves the repair of damaged or diseased organs or tissues.

14. Give examples of the therapeutic use of stem cells such as corneal repair and the regeneration of damaged skin.

15. State that stem cells from an embryo can self-renew, under the right conditions, *in vitro*.

16. Describe the research uses of stem cells such as model cells to study how diseases develop and for drug testing.

17. Explain that stem cell research provides information on how cell processes such as cell growth, differentiation and gene regulation work.

18. Explain that ethical issues arise around the use of embryonic stem cells to offer effective treatments for disease and injury because it involves destruction of embryos.

19. State that cancer cells divide excessively because they do not respond to regulatory signals, resulting in a mass of abnormal cells called a tumour.

20. Describe cells within the tumour failing to attach to each other and spreading through the body where they may form secondary tumours.

Skills

1. *Select information from a double axis line graph.*

2. *Process information selected from a double axis line graph.*

3. *Predict from a double axis line graph.*

1.2 Structure and replication of DNA

You should already know:

- The nucleus of a cell contains chromosomes, which are composed of genes made of DNA.
- Chromosomes are replicated prior to cell division.
- DNA carries genetic information for making proteins.
- DNA molecules are in the shape of a double-stranded helix held together by complementary base pairs.
- The four bases in DNA are called adenine (A), guanine (G), thymine (T) and cytosine (C).
- In DNA, A is always paired with T; C is always paired with G.
- The sequence of bases along one of the DNA strands makes up the genetic code.
- A gene is a section of DNA which codes for a protein.

Learning intentions

- Be familiar with the use of gel electrophoresis in the separation of macromolecules such as DNA fragments.
- Describe the structure of a DNA molecule in terms of shape, nucleotides, sugar–phosphate backbone, antiparallel strands and complementary base pairing.
- Describe the roles of primers, DNA polymerase and ligase in DNA replication.
- Explain directionality of replication on both template strands of DNA.
- Explain the stages and temperatures in the amplification of DNA by the polymerase chain reaction (PCR).
- Describe practical applications of PCR.

Introducing DNA

Deoxyribonucleic acid (DNA) is the substance that makes up the genetic material in human cells. The chemical structure of DNA was discovered in the 1950s by two scientists, James Watson and Francis Crick, working in Cambridge, UK. With Maurice Wilkins, they won the 1962 Nobel Prize in Physiology or Medicine for their discoveries. The evidence for their conclusions came from model-building using X-ray images of DNA crystals from the work of Rosalind Franklin and Maurice Wilkins, and others, as shown in **Figure 1.2.1**.

DNA carries the genetic code which determines and regulates how and when to make proteins.

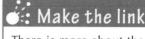 Make the link

There is more about the genetic code in Chapter 1.3 on page 35 and Chapter 1.5 on page 54.

(a)

(b)

Figure 1.2.1 *(a) James Watson (left) and Francis Crick with their model of DNA (b) Rosalind Franklin whose X-ray images of DNA informed the modelling*

Structure of DNA

Nucleotides

DNA is composed of very long molecules made up of repeating chemical units called **nucleotides**.

A nucleotide has three component parts as shown in **Figure 1.2.2**:

- a nucleotide **base**
- a phosphate group
- a central 5-carbon **deoxyribose** sugar

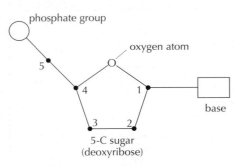

Figure 1.2.2 *A single DNA nucleotide*

Notice that the carbon atoms of the deoxyribose sugar are numbered. The base is joined to carbon 1 and the phosphate group to carbon 5.

There are four different nucleotide bases called adenine (A), guanine (G), thymine (T) and cytosine (C). Nucleotides are linked together to make a long strand of DNA. This arrangement produces a sugar–phosphate backbone running in a **3′–5′** direction with the bases attached as shown in **Figure 1.2.3**. It is the base sequence in DNA which forms the genetic code. A deoxyribose sugar is found at the 3′ end of the strand and a phosphate at the 5′ end.

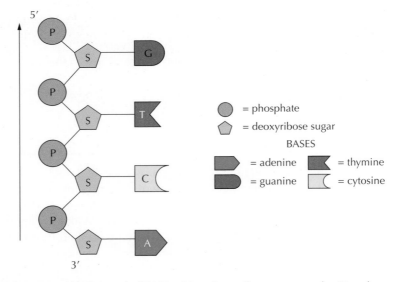

Figure 1.2.3 *A short strand of DNA with a deoxyribose sugar at the 3′ end and a phosphate at the 5′ end of the sugar–phosphate backbone*

Complementary base pairing

Two strands of DNA are joined together through weak hydrogen bonds that link **complementary** pairs of nucleotide bases. Adenine on one strand always pairs with thymine on the other and guanine on one strand always pairs with cytosine on the other. The strands run from carbon 3′ (prime) at one end to carbon 5′ (prime) at the other end but in opposite directions and are described as **antiparallel**, as shown in **Figure 1.2.4**. Deoxyribose is at the 3′ end of a strand and phosphate is at the 5′ end.

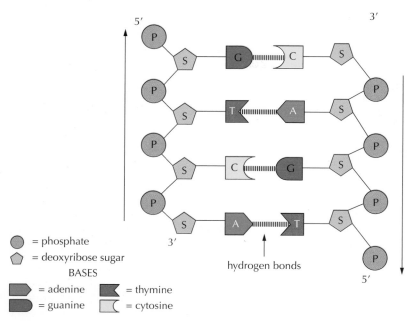

● = phosphate
⬠ = deoxyribose sugar
          BASES
◗ = adenine   ◖ = thymine
◗ = guanine   ◖ = cytosine

Figure 1.2.4 *A short section of DNA to show complementary base pairing and the antiparallel structure of the double strands*

Double helix

Long sections of DNA are twisted into the shape of a **double-stranded helix** as shown in **Figure 1.2.5**. The hydrogen bonds which link the bases of each strand are weak and can be broken and re-joined easily.

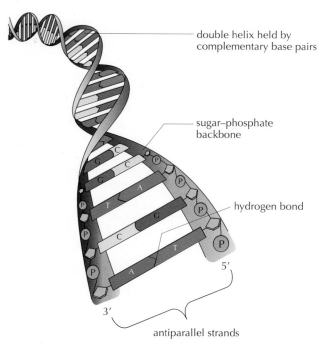

double helix held by complementary base pairs

sugar–phosphate backbone

hydrogen bond

5′

3′

antiparallel strands

Figure 1.2.5 *A long section of DNA showing the sugar–phosphate backbones wound into a double-stranded helix shape*

🔍 **Hint**

Remember to use the term 'complementary' when describing base pairs – it's the best and most meaningful way of talking about them.

📖 **Antiparallel**

The two strands of a DNA molecule run in opposite directions and are said to be antiparallel to each other. One strand has deoxyribose (3′) at one end of the molecule, but its complementary strand has a phosphate group (5′) at the same end of the molecule.

🔍 **Hint**

Antiparallel is a tricky idea – remember the strands are like the two lanes of traffic travelling different ways on a road; the same but going in the opposite direction.

DNA replication

It is essential that DNA molecules can replicate prior to cell division to make exact copies of chromosomes so that human cells have enough genetic information to produce two genetically identical daughter cells. The basis for the replication process is the complementary base pairing mentioned above:

- adenine pairs with thymine
- cytosine pairs with guanine

For DNA replication to occur, the following substances must be present within the nucleus of a cell which is going to divide:

- DNA templates – the parental DNA strands
- free DNA nucleotide with all four base types
- the enzymes **DNA polymerase** and **ligase**
- **primers** to start the replication process
- adenosine triphosphate (ATP) to supply energy for the process

Events in DNA replication

A parental DNA double helix unwinds and hydrogen bonds between bases break to separate the two parental DNA templates at the replication fork. Primers bind to the 3′ ends of each template strand.

DNA polymerase adds free DNA nucleotides to the primers using complementary base pairing. The **leading strand** of DNA is replicated continuously in a 3′ to 5′ direction. The **lagging strand** is replicated in fragments: as the parental strand continues to unwind and hydrogen bonds break, 3′ sites become available in the replication fork. The fragments are then joined by DNA ligase. Two genetically identical daughter DNA molecules are formed which are also identical to the parental DNA because they contain exactly the same sequence of bases. The process of DNA replication is shown in **Figure 1.2.6**.

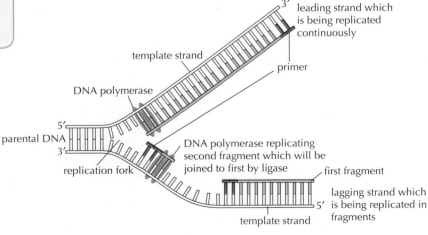

Figure 1.2.6 *Replication of parental DNA to produce two identical daughter copies of the parental DNA*

Polymerase chain reaction

The **polymerase chain reaction (PCR)** is a technique for the amplification of DNA *in vitro*. This technique is used to produce billions of copies of a sample of DNA which would otherwise be too small to be of use for the various applications outlined below. The process is carried out in small tubes placed in a series of water baths or, more usually, in a machine called a thermal cycler as shown in **Figure 1.2.7**.

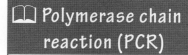

📖 **Polymerase chain reaction (PCR)**

A laboratory *in vitro* technique for the amplification of DNA.

(a)

(b)

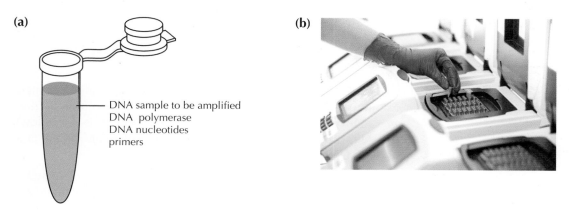

Figure 1.2.7 *(a) Contents of a PCR tube (b) Photograph of a thermal cycler machine*

The DNA to be amplified is placed in the tube with primers which are complementary to specific target sequences at the two 3′ ends of the DNA. Heat-tolerant DNA polymerase adds complementary nucleotides to the ends of both template strands at once. The tube is cycled through a series of temperatures and each cycle doubles the number of copies of the DNA being amplified as shown in **Figure 1.2.8(a)**. **Figure 1.2.8(b)** shows a graph of the changes in temperature of one thermal cycle.

📖 *In vitro*

Carried out in containers in a laboratory rather than in a living system.

(a)

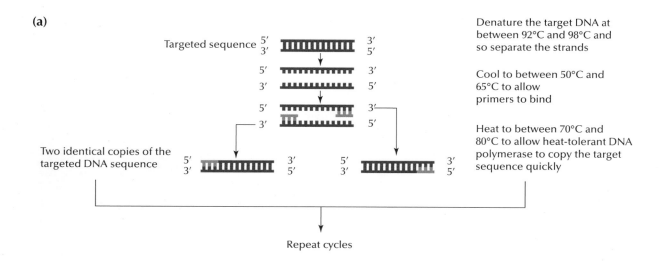

Targeted sequence 5′/3′

Denature the target DNA at between 92°C and 98°C and so separate the strands

Cool to between 50°C and 65°C to allow primers to bind

Heat to between 70°C and 80°C to allow heat-tolerant DNA polymerase to copy the target sequence quickly

Two identical copies of the targeted DNA sequence

Repeat cycles

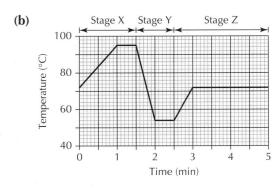

Figure 1.2.8 *(a) Description of stages in PCR (b) Graph to show the thermal changes in a cycle of PCR: Stage X in which the strands are separated, Stage Y when primers bind and Stage Z when heat-tolerant DNA polymerase replicates the strands*

Applications of PCR

PCR amplifies DNA, which can then be used for a number of applications. PCR can help solve crimes by amplifying tiny amounts of DNA from biological samples such as blood or semen from a crime scene so that they can be analysed by gel electrophoresis (see below). The results can be used to confirm the presence of an individual, or at least their DNA, at the scene of a crime.

In paternity cases, PCR can provide quantities of DNA to be used to determine the biological father of a child. This can be indicated through the comparison of his DNA with that of the child and can help settle paternity suits in which the identity of the father of a child needs to be confirmed.

In medicine, the diagnosis of genetic disorders requires lots of tests and in turn these tests require a lot of DNA, which can be produced using PCR. This allows test results to be as reliable as possible.

⚙ Gel electrophoresis

📖 Gel electrophoresis

Laboratory *in vitro* technique for the separation of macromolecules such as DNA fragments to produce unique DNA profiles.

Once DNA has been amplified by PCR it can be treated with enzymes to produce specific fragments and then the fragments can be stained ready for separation by **gel electrophoresis**. A sample of the stained fragments is placed in a well cut into a block of agarose gel. An electric current can be passed through the gel and the stained fragments move in the gel depending on their size and electrical charge. The result is a profile of bands in the gel which can be further stained to make them more visible. The profile is unique to the individual whose DNA sample is being tested.

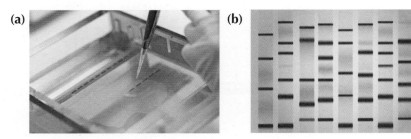

Figure 1.2.9 *(a) Photograph of a gel electrophoresis technique in which stained DNA fragments are being dropped into wells in a block of agarose gel (b) DNA profiles produced by gel electrophoresis – each 'ladder' or profile is unique to the individual whose DNA is being analysed*

Applications of gel electrophoresis

Figure 1.2.10 shows a gel electrophoresis DNA profile taken from a crime scene. Also shown are DNA profiles from the victim of the crime and from two individual suspects. As can be seen, the profile from the crime scene matches that of Individual 1, which makes it certain that this individual was present at the scene – unless they have an identical twin or their DNA was planted there!

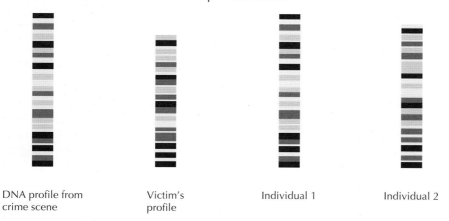

DNA profile from crime scene   Victim's profile   Individual 1   Individual 2

Figure 1.2.10 *Use of gel electrophoresis profile in solving crimes*

GO! Activity 1.2.1 Work individually to . . .

Structured questions

1. DNA is the substance which carries the human genetic code.
   a) **Describe** the shape of a DNA molecule. 1
   b) **Explain** the role of hydrogen bonds in DNA structure. 1
   c) **Name** the components of a single nucleotide. 1
   d) **Describe** how the following terms are applied to the structure of a DNA molecule:
      i. Complementary base pairing 1
      ii. Antiparallel strands 1

2. The stages in a polymerase chain reaction (PCR) are shown below.
   Stage 1: DNA heated to 92°C
   Stage 2: DNA cooled to 60°C
   Stage 3: DNA heated to 70°C
   a) **Describe** how the temperature used in **Stage 1** affects the structure of DNA. 1
   b) **Give** the main event that occurs during **Stage 2**. 1
   c) **Describe** the role of DNA polymerase in **Stage 3**. 1
   d) **Calculate** the number of cycles of PCR which would be required to produce 64 copies of a single DNA molecule. 1
   e) **Give one** application of PCR. 1

3. The table on the right is based on results obtained in 1952 by the scientist Erwin Chargaff who investigated the proportions of bases in DNA samples from cells from two human sources.

   a) **Give** evidence that adenine pairs with thymine and cytosine pairs with guanine. 1

   b) **Give** evidence that some experimental error affected these results. 1

Bases	Concentration (units)	
	Sperm	Liver
adenine	0·29	0·27
guanine	0·18	0·19
thymine	0·30	0·27
cytosine	0·18	0·18

4. When a DNA sample is heated, its strands separate. The heat needed to do this is proportional to the percentage of G–C base pairs as shown in the table below.

G–C base pairs (% in molecule)	Temperature needed to separate strands (°C)
0	70
20	77
40	85
60	94
80	100

On a piece of graph paper, plot the percentage of G–C base pairs against the temperature at which the strands of DNA molecules separate. 2

Make the link

There is information about drawing line graphs in Chapter 4 on page 266.

5. Patients requiring an organ transplant are tissue typed to match with potential donors.

In a procedure, amplified DNA from a patient and three potential donors was compared by gel electrophoresis as shown in the diagram. A DNA ladder with fragments of known size was also run in the same gel for reference. The table shows how the size of a DNA fragment affects the distance it travels in the gel.

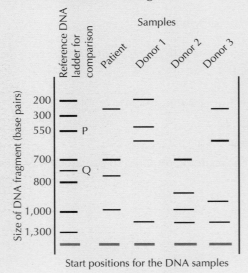

Size of DNA fragment (bp)	Distance travelled (mm)
200	70
300	56
550	30
700	16
800	10
1,000	8
1,300	6

(continued)

a) **Identify two** variables which must be controlled when carrying out electrophoresis. 2
b) **Suggest** what should be done to increase the reliability of the procedure. 1
c) **Identify** the distance travelled by fragment P. 1
d) On a piece of graph paper, draw a line graph to show that the size of a DNA fragment affects how far it travels in the gel. 2
e) **Describe** the relationship between the size of a DNA fragment and the distance it travels in the gel. 1
f) Fragment Q has travelled 13 mm in the gel.

   Estimate the number of base pairs in this fragment. 1
g) **Identify** the Donor whose DNA best matches that of the Patient. 1

 ## Make the link

There is information about variables and reliability in Chapter 4 on page 263.

Extended response questions

1. **Give an account** of the role of enzymes in DNA replication. 4
2. **Give an account** of the replication of DNA. 6

GO! Activity 1.2.2 Work in pairs to …

1. **Practical activity: Isolation of DNA**

You will need: suitable fruit, e.g. strawberries or kiwi fruit, plastic bag, measuring cylinder, water, salt, detergent, rubber band, water bath at 60°C, ice bath, boiling tube, test tube rack, filter funnel, filter paper and ethanol.

⚠ Your teacher will give specific safety instructions for carrying out this experiment.

- Be careful not to spill from the bag.
- Remember that the water bath is hot – take care not to drip hot water.

Method

- Put your strawberry or kiwi fruit into the plastic bag. Hold the bag closed and squeeze the fruit to a pulp.
- Make an extraction fluid by combining the following: 90 cm^3 of water, 5 g of salt and 10 cm^3 of detergent.
- Add 30 cm^3 of the extraction fluid into the bag and mix it with the pulped fruit for 1 minute.
- Close the bag with a rubber band and put it in the water bath for 15 minutes.
- Place the bag in the ice bath for 5 minutes.
- Filter the liquid in the bag to remove any remaining solid fruit.
- Half fill a boiling tube with the filtered liquid.

DNA strands appearing

Tilt the boiling tube, then carefully and slowly pour ice-cold ethanol down the side of the boiling tube. The ethanol will float on top of the liquid – you should see the fruit DNA appearing in the ethanol layers as shown in the photo – it shows DNA extracted from a strawberry.

2. **Practical activity: The DNA Detective**

   a) *You will need: You will need internet access to go to* https://www2.le.ac.uk/projects/vgec/schoolsandcolleges/dnadetective *and download and print the PDFs "Crime scene report" and "DNA profiles". Your teacher may have already done this for you.*

   Read the Crime Scene Report and do the task on page 4. **DO NOT** read the answers which are on page 5 until you have finished the task!

   b) Discuss the following questions with your partner. Your teacher may organise a class discussion of these and other issues. Be ready to take part.

   i. Do you think that using DNA profiling to solve crimes is a good thing?

   ii. Should there be a national database of DNA profiles?

   iii. Should everyone in the UK have their DNA profile on record?

   iv. Does a DNA profile found at a crime scene mean that the crime has been solved?

   v. In what ways might an innocent person's DNA profile appear at a crime scene?

3. **Card sequencing activity: Sequencing event in DNA replication**

   You will need: a stopwatch and a mini whiteboard with a marker pen.

   • Read the phrases in the grid below which describe events in DNA replication.

   • Work together to arrange the phrases in the correct order – start with phrase 8. Mark your answer on the whiteboard. Your teacher will check your work or put the correct sequence on the board.

   • You should each try this again individually against the clock – who is faster to the correct answer?

   • Your teacher may provide you with a photocopy of the grid. If so, cut it into strips, re-sequence it and glue it into your notes to make a permanent flowchart.

1 Primer binds to the 3' end of the leading strand.	5 The fragments on the lagging strand are joined by the enzyme DNA ligase.
2 Primers bind to 3' deoxyribose of the lagging strand at intervals as they become exposed by base pair breakages.	6 DNA polymerase adds complementary DNA nucleotides to the exposed 3' ends of the lagging strand in fragments.
3 DNA polymerase adds complementary DNA nucleotides to the deoxyribose (3') end of the leading strand continuously.	7 Weak hydrogen bonds between complementary base pairs on each strand break.
4 Two new and identical double helices are formed.	8 DNA uncoils from one end.

4. **Flashcard activity**

   You will each need: a set of blank flashcards (A7 cards) and a stopwatch.

   • Find the glossary terms for this chapter – they are the **black** typeface and **red** typeface terms. Using your blank cards, you should each make a set of flashcards for these terms – write the term on one side and the definition on the other. You will find the definitions in the chapter.

(continued)

- Shuffle your cards and lay them out in a column, some showing terms and some showing definitions – you decide. Your partner should match their cards with yours, laying their cards in a column beside yours to give the corresponding term or definition. Time how long they take to do this.
- Now swap roles – your partner should lay out their cards and you should try the matching exercise while your partner times you.
- You should each keep your set of flashcards as a revision tool for later.

GO! Activity 1.2.3 Work as a group to ...

1. **Design and make a model of DNA.**

   You will need: two different colours of miniature marshmallow sweets to represent sugars and phosphates, four colours of jelly sweets for the bases and cocktail sticks cut in two to join the components.

   Arrange five nucleotide pairs like a ladder – don't try to model the double helix!

2. **Design and make an A2 collage to show the structure of DNA.**

   You will need: an A2 sheet, six pieces of different coloured card, scissors, a glue stick and a marker pen.

   Work together to make a collage entitled 'The structure of DNA'. Make sure you label the parts of your collage.

   Your teacher may ask your group to present your work to the class.

3. **Research the work of Matthew Meselson and Frank Stahl on the replication of DNA from the 1950s.**

   Prepare a set of three PowerPoint slides to explain Meselson and Stahl's research. You could use the diagram in this link to help you:

   www.nature.com/scitable/content/the-meselson-stahl-experiment-18551/

   Ensure you mention their aim, the importance of the heavy isotope of nitrogen they used and the conclusion they were able to reach. Be prepared to present your work to the class.

Learning checklist

After working on this chapter, I can:

Knowledge and understanding

1. State that DNA carries the genetic code.

2. State that DNA molecules are double-stranded helices made up of nucleotides.

3. State that the strands of DNA are antiparallel and linked through hydrogen bonds through complementary base pairs.

4. Describe antiparallel in terms of 3' and 5' ends of DNA.

5. Describe each nucleotide as being composed of deoxyribose sugar, a phosphate and a base.

6. Name the four types of nucleotide base in DNA as adenine, guanine, thymine and cytosine.

7. State that the complementary base pairs are adenine with thymine and guanine with cytosine.

8. Explain that DNA must replicate prior to cell division so that daughter cells can each inherit an exact copy of the genetic information in the parent cell.

9. Describe replication in terms of the unwinding of parental DNA, breaking of hydrogen bonds, binding of primers and the action of DNA polymerase and ligase.

10. State that DNA polymerase adds free DNA nucleotides to template strands of DNA in a 3' to 5' direction.

11. State that the leading strand is replicated continuously and the lagging strand is replicated in fragments.

12. State that DNA ligase joins fragments of DNA on the lagging strand.

13. State that the polymerase chain reaction (PCR) is an *in vitro* method of amplifying DNA samples.

14. Describe the thermal cycle of PCR as between 92°C and 98°C to separate strands, between 50°C and 65°C to allow primers to bind and between 70°C and 80°C to speed up the action of DNA polymerase.

15. Describe gel electrophoresis as a technique used to separate macromolecules, such as fragments of DNA, in an electric field.

16. State that a DNA profile produced by gel electrophoresis is unique to an individual.

Skills

1. *Process information to calculate numbers of DNA molecules produced in PCR.*

2. *Select information from a table and a graph.*

3. *Present data as a line graph.*

4. *Identify variables in the planning of a procedure.*

5. *Evaluate an experiment to improve reliability.*

1.3 Gene expression

The genetic code in DNA

The genetic code of humans helps determine what we look like and how our bodies function. The sequence of bases along strands of DNA makes up the code. The DNA encodes instructions for the production of specific human proteins which cells can assemble using amino acids from the diet. **Figure 1.3.1** shows a selection of foods eaten by humans, from both plants and animals, which are rich in protein and can be digested to produce the supply of amino acids needed for protein synthesis.

The genetic code in DNA is read in groups of three bases called **codons**. Each codon encodes a specific amino acid. This sequence is often called the **genome**. It is the protein produced by the expression of genes in the genome, as well as the influence of environmental factors such as diet, which determines the **phenotype** of the individual human as shown in **Figures 1.3.2** and **1.3.3**.

Figure 1.3.1 *A selection of protein-containing foods – it is from foods such as these that our amino acid supplies come!*

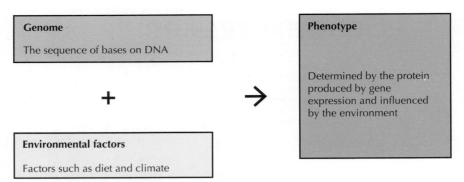

Genome	
The sequence of bases on DNA	

+

Environmental factors
Factors such as diet and climate

→

Phenotype
Determined by the protein produced by gene expression and influenced by the environment

Figure 1.3.2 *The factors which contribute to the phenotype of a human*

📖 **Codon**

A sequence of three bases on DNA and mRNA which codes for a specific amino acid.

📖 **Phenotype**

The observable characteristics of a human, including aspects such as blood group which must be tested for.

Figure 1.3.3 *Human variation – our different phenotypes are determined by our genes but the environment also plays a role. Can you think of examples of environmental influences on phenotype?*

Figure 1.3.4 *A small section of an RNA strand*

ribose sugar

A

G

U

uracil base

C

Introducing RNA

During gene expression, the base sequences on DNA are copied to form complementary molecules of a substance called ribonucleic acid (RNA). Some of the RNA molecules carry complementary copies of the DNA codes which can be translated to produce protein from amino acids. Other RNA molecules are involved in the processes which actually assemble the proteins.

RNA is similar to DNA and, like DNA, is made up from strands of nucleotides. RNA nucleotides differ from those of DNA because the sugar component is a sugar called **ribose** rather than deoxyribose. The bases found in RNA include adenine (A), guanine (G) and cytosine (C) but there is no thymine and another base called **uracil (U)** is found in its place. RNA molecules can be very long but they are always single stranded as shown in **Figure 1.3.4**.

Types of RNA

There are three main types of RNA called **messenger RNA** (mRNA), **transfer RNA** (tRNA) and **ribosomal RNA** (rRNA) as shown in **Figure 1.3.5**.

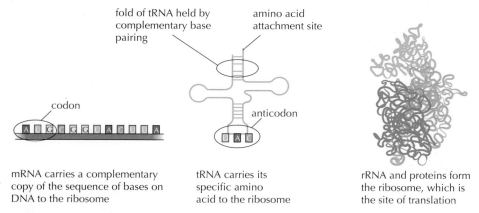

mRNA carries a complementary copy of the sequence of bases on DNA to the ribosome

tRNA carries its specific amino acid to the ribosome

rRNA and proteins form the ribosome, which is the site of translation

Figure 1.3.5 *The structures and functions of messenger RNA (mRNA), transfer RNA (tRNA) and ribosomal RNA (rRNA)*

Only a fraction of the genes in a particular cell are expressed. Gene expression involves two stages called **transcription** and **translation**. The different RNA types have individual functions in these stages.

Transcription

In the nucleus of a cell, DNA is transcribed to produce the various forms of RNA.

Primary transcripts and splicing

During transcription of a gene, sequences of DNA are used to form the templates for the production of a **primary transcript** of DNA. At one end of the gene is a **start codon,** which indicates where transcription should start, and at the other end is a **stop codon,** which indicates where transcription should be terminated. The enzyme **RNA polymerase** unwinds the double helix and breaks hydrogen bonds between complementary base pairs in the DNA within the gene and then adds complementary RNA nucleotides against the exposed DNA bases. The adjacent RNA nucleotides are linked together to form a primary transcript as shown in **Figure 1.3.6**. The primary transcript is made up of coding regions called **exons**, which code for amino acids making up protein, and non-coding regions called **introns**, which do not. The primary transcript undergoes **RNA splicing** in which introns are removed and the remaining exons are spliced together to produce a mature mRNA molecule. The order of the exons does not change during splicing.

📖 Transcription

The copying of DNA sequences in the nucleus to produce a primary transcript.

📖 Translation

The production of a polypeptide at a ribosome using the DNA sequences encoded in mRNA.

📖 RNA polymerase

Enzyme which unwinds the double helix and breaks hydrogen bonds between complementary base pairs in the DNA and adds nucleotides to produce a primary transcript.

📖 RNA splicing

The removal of introns from a primary transcript.

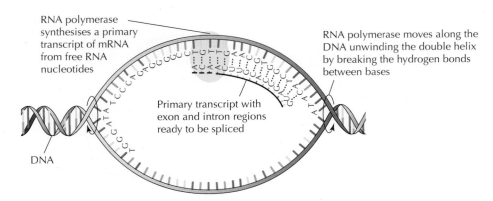

RNA polymerase synthesises a primary transcript of mRNA from free RNA nucleotides

RNA polymerase moves along the DNA unwinding the double helix by breaking the hydrogen bonds between bases

Primary transcript with exon and intron regions ready to be spliced

DNA

Figure 1.3.6 *Transcription of a gene by RNA polymerase*

Alternative RNA splicing

A transcript from a single gene can be alternatively spliced by the treatment of certain exons as introns, removing them to produce a range of different mRNAs as shown in **Figure 1.3.7**. By this means, a cell can use one gene to produce different proteins depending on its age, metabolic needs, health and other factors. Note that in the mature transcripts, the remaining exons are always in the same order.

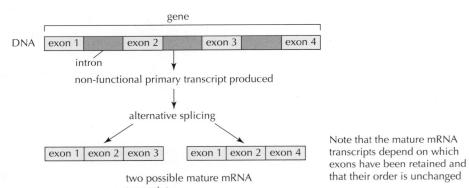

gene

DNA | exon 1 | exon 2 | exon 3 | exon 4

intron

non-functional primary transcript produced

alternative splicing

exon 1 | exon 2 | exon 3   exon 1 | exon 2 | exon 4

two possible mature mRNA transcripts

Note that the mature mRNA transcripts depend on which exons have been retained and that their order is unchanged

Figure 1.3.7 *Alternative RNA splicing*

Translation

Mature messenger RNA (mRNA) carries its complementary copy of the DNA code, including start and stop codons, from the nucleus to a ribosome where it attaches.

Transfer RNA (tRNA) has a folded shape due to complementary base pairing and carries a specific amino acid bound to its amino acid attachment site as shown in **Figure 1.3.5**. Each different type of tRNA carries a specific **anticodon** made up of three bases which allows it to bind to a specific codon on the mature mRNA. tRNA molecules with an amino acid attached approach the ribosome, where their anticodons align against the complementary codons of mRNA. This brings their amino acids into alignment and encourages the formation of **peptide bonds** between adjacent amino acids as shown in **Figure 1.3.8**. The **polypeptide** formed can fold to make the required protein.

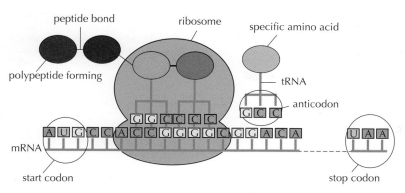

Figure 1.3.8 *Translation at a ribosome – amino acids are lined up according to the base sequence on mRNA and then bonded together with peptide bonds*

Protein structure

As described above, amino acids are linked by strong peptide bonds to produce polypeptides. Polypeptide chains fold to form the three-dimensional shape of a protein. The highly specific shape of an individual protein molecule is held in place by hydrogen bonds and other interactions between individual amino acids. **Figure 1.3.9** shows an image of a protein showing its complex and highly specific three-dimensional shape.

Protein functions

The three-dimensional shape of an individual protein as shown in **Figure 1.3.9** determines its function. The table below shows some of the functions of proteins studied in the Higher Human Biology course.

Make the link

There is more about PCR in Chapter 1.2 on page 27.

Figure 1.3.9 *A three-dimensional impression of the shape of a protein – this is the enzyme heat-tolerant DNA polymerase involved in PCR*

Make the link

There is more about muscle fibres in Chapter 1.8 on page 84, insulin and glucagon in Chapter 2.8 on page 175, antibodies in Chapter 3.6 on page 231, insulin receptors in Chapter 2.8 on page 177, DNA and RNA polymerase and ligase in Chapter 1.2 on page 26, ATP synthase in Chapter 1.7 on page 75 and myoglobin in Chapter 1.8 on page 84.

Type of protein	Function	Examples
Structural	Form the structure of the cells and tissues	Slow- and fast-twitch fibres in muscle tissues contain proteins which are important in body movements
Peptide hormones	Carry chemical messages in the bloodstream	Insulin and glucagon help regulate blood glucose levels
Antibodies	Fight infection as part of the immune response	Flu antibodies specifically recognise and bind to flu antigens to trigger immune responses
Membrane receptors	Are sensitive to signal molecules such as hormones	Liver cell receptors are sensitive to insulin
Enzymes	Catalyse chemical reactions in cells	DNA and RNA polymerases, ligase, ATP synthase
Transporters	Act as carriers of other substances	Haemoglobin and myoglobin carry oxygen; the electron transport chain proteins carry electrons

GO! Activity 1.3.1 Work individually to ...

Structured questions

1. RNA is a substance which is very important in gene expression.
   a) **Give two** differences between the structure of DNA and that of RNA.   2
   b) **Give** the meaning of the term 'codon'.   1
   c) **State** what is meant by a start and a stop codon.   2
   d) **Explain** the difference between an intron and an exon.   2

2. Proteins are polypeptide chains folded into different shapes.
   a) **Name** the strong bonds which hold amino acids together in sequence along a polypeptide chain.   1
   b) **Describe** how polypeptides are held in the three-dimensional shape of proteins.   1
   c) **Explain** the importance of the three-dimensional shape of a protein.   1

3. The table shows the positions of bases in each mRNA codon and the amino acid they encode in humans.

First position	Second position				Third position
	U	C	A	G	
U	phenylalanine	serine	tyrosine	cysteine	U
					C
	leucine		stop	stop	A
			stop	tryptophan	G
C	leucine	proline	histidine	arginine	U
					C
			glutamine		A
					G
A	isoleucine	threonine	asparagine	serine	U
					C
			lysine	arginine	A
	start/methionine				G
G	valine	alanine	aspartic acid	glycine	U
					C
			glutamic acid		A
					G

   a) **Give** the mRNA codon for the amino acid methionine.   1
   b) **Identify** the number of different amino acids which have the base G as the second base in the codon.   1
   c) **Give one** DNA sequence which would encode the following three amino acids:
      --- isoleucine – cysteine – tryptophan ---   1
   d) **Name** the amino acid which could be carried by a tRNA molecule with the anticodon CAU.   1

Extended response questions

1. **Give an account** of the production of messenger RNA under the following headings:
   a) Transcription of DNA   5
   b) Splicing of RNA   2
2. **Give an account** of the translation of mRNA.   5

Activity 1.3.2 Work in pairs to …

1. **Make a model of the genetic code.**

   You will need: a pink poppet bead, a supply of red, green, blue and yellow beads (or modelling clay/dough), a few strips of white paper and a marker pen.

   The table below shows some imaginary genetic code for four amino acids.

   The code is made by joining together some poppet beads.

Amino acid	P	Q	R	S
DNA code	red–green–blue	yellow–red–green	blue–yellow–red	green–blue–yellow

   - Mark out a strip of paper as shown below and add your choice of amino acid letters from the table to the ovals to make an imaginary protein like this:

   - Make up a DNA molecule with poppet beads. Use the pink bead as the start point and add coloured beads to make a code for your imaginary protein using information in the table.
   - Swap your DNA with another pair and try to de-code their DNA on another paper strip.
   - Repeat a few more times.
   - Now try the other way round: invent a protein on a strip and see if another group can make the correct DNA model.

2. **Design and make an A3 collage to show protein function.**

   You will need: a piece of A3 card, coloured paper, marker pens, scissors and a glue stick or sticky tape.

   Work together to make a collage entitled 'Protein function in humans'. Divide the card into six sections and use the information in the table on page 39 to present some information about the human proteins mentioned. You could each choose three proteins to present.

   Ensure you name each protein type and illustrate what each does in life.

3. **Card sequencing activity: Transcription and translation**

   - Read the phrases in the grid below which describe events in transcription and translation.
   - Decide if statements 1–12 are related to transcription or translation and make a list of each. Ask your teacher to check your lists.
   - Arrange the numbers in each list to show the sequences of events in transcription of DNA and translation of mRNA. Ask your teacher to check your sequences.
   - Your teacher may provide you with a photocopy of the grid. If so, cut it into strips, re-sequence it and glue it into your notes to make a permanent flowchart.

1 Mature mRNA attaches to a ribosome	7 RNA polymerase unwinds DNA and breaks hydrogen bonds between the two strands
2 Peptide bonds form between amino acids and a polypeptide forms	8 Introns are removed from the primary transcript
3 RNA polymerase aligns free RNA nucleotides with their complementary nucleotides on the DNA template strand	9 The template strand is separated from its complementary strand
4 A primary transcript containing exon (coding) and intron (non-coding) regions is produced	10 tRNA carries specific amino acid to the mRNA at a ribosome
5 Amino acids are aligned into sequence	11 Anticodons on tRNA pair with their complementary codons on mRNA
6 tRNAs bind to their specific amino acids in cytoplasm	12 Exons are spliced together to form mature mRNA

 4. Flashcard activity

You will each need: a set of blank flashcards (A7 cards) and a stopwatch.

- Find the glossary terms for this chapter – they are the **black** typeface and **red** typeface terms. Using your blank cards, you should each make a set of flashcards for these terms – write the term on one side and the definition on the other. You will find the definitions in the chapter.

- Shuffle your cards and lay them out in a column, some showing terms and some showing definitions – you decide. Your partner should match their cards with yours, laying their cards in a column beside yours to give the corresponding term or definition. Time how long they take to do this.

- Now swap roles – your partner should lay out their cards and you should try the matching exercise while your partner times you.

- You should each keep your set of flashcards as a revision tool for later.

Activity 1.3.3 Work as a group to ...

1. Research the Human Genome Project.

Visit the web page below:

www.genome.gov/human-genome-project

Split your group into three to work on the following sub-topics:

- What is the Human Genome Project?
- Human Genome Project results
- Human Genome Project timeline of events

Each group should produce a single PowerPoint slide or storyboard to illustrate their findings on their sub-topic.

Combine the slides or storyboards to form a presentation or display and be prepared to present it to your class.

Learning checklist

After working on this chapter, I can:

Knowledge and understanding

1. State that genes are the units of the genetic code that make up the human genome.

2. Explain that genes are expressed to produce proteins, which form the structure and control the functions of the human body.

3. State that only a fraction of the genes in a human cell are expressed.

4. State that gene expression involves the transcription and translation of DNA sequences.

5. State that transcription and translation involve three types of RNA (mRNA, tRNA and rRNA).

6. State that RNA is single-stranded and is composed of nucleotides containing ribose sugar, phosphate and one of four bases: cytosine, guanine, adenine and uracil.

7. State that messenger RNA (mRNA) carries a copy of the DNA code from the nucleus to the ribosome.

8. State that mRNA is transcribed from DNA in the nucleus and translated into proteins by ribosomes in the cytoplasm.

9. State that each triplet of bases on the mRNA molecule is called a codon and codes for a specific amino acid.

10. State that transfer RNA (tRNA) folds due to complementary base pairing.

11. State that each tRNA molecule carries its specific amino acid to the ribosome.

12. State that a tRNA molecule has an anticodon (an exposed triplet of bases) at one end and an attachment site for a specific amino acid at the other end.

13. State that ribosomal RNA (rRNA) and proteins are components of ribosomes.

14. State that DNA in the nucleus is transcribed to produce messenger RNA (mRNA), which carries a copy of the genetic code.

15. State that in transcription, the enzyme RNA polymerase moves along DNA unwinding the double helix and breaking the hydrogen bonds between the bases.

16. State that RNA polymerase synthesises a primary transcript of mRNA from RNA nucleotides by complementary base pairing.

17. State that uracil in RNA is complementary to adenine.

18. State that human genes have introns (non-coding regions) and exons (coding regions).

19. State that the introns of the primary transcript are non-coding regions and are removed.

20. State that the exons are coding regions and are joined together to form the mature transcript.

21. State that the order of the exons is unchanged during splicing.

22. State that different proteins can be expressed from one gene, as a result of alternative RNA splicing.

23. Explain that different mature mRNA transcripts are produced from the same primary transcript depending on which exons are retained.

24. State that tRNA is involved in the translation of mRNA into a polypeptide at a ribosome.

25. State that translation begins at a start codon and ends at a stop codon.

26. State that amino acids are carried by specific tRNA molecules.

27. State that tRNA anticodons align with and bind to mRNA codons by complementary base pairing, translating the genetic code into a sequence of amino acids.

28. State that amino acids are linked by peptide bonds to form polypeptides.

29. State that polypeptide chains fold to form the three-dimensional shape of a protein, held together by hydrogen bonds and other interactions between individual amino acids.

30. State that proteins have a large variety of shapes which determine their functions.

31. State that the phenotype of an individual is determined by proteins produced as the result of gene expression.

32. State that environmental factors also influence phenotype.

Skills

1. *Select information from a table.*

2. *Process information about codons and anticodons.*

1.4 Mutation

You should already know:

- A mutation is a random change to genetic material.

Learning intentions

- Describe the different types of gene mutation.
- Describe the effects of mutation in terms of proteins expressed.
- Describe the different types of chromosome structure mutation.

Introducing mutation

Mutations are changes in the DNA that can result in no protein or an altered protein being synthesised. These changes happen randomly and rarely but their frequency can be increased by environmental factors, including radiation and some chemicals.

Cancer

A mutation arising in a somatic cell can cause the cell to become cancerous, dividing excessively to form a tumour. Exposure of cells to radiation and certain chemicals can increase the likelihood of this happening, which has led to health concerns. For example, the effects of ultraviolet (UV) light on skin cells and the effects of various chemicals in tobacco on lung cells are well known and have led to high-profile advice on the use of sunblock and the avoidance of tobacco as shown in **Figure 1.4.1**.

Inherited mutations and genetic conditions

In cases where a mutation has affected germline cells, the condition it causes can be inherited in families. These conditions include sickle-cell disease, cystic fibrosis and haemophilia.

(a) 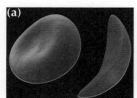 **(b)**  **(c)**

Figure 1.4.2 *(a) In sickle-cell disease, red blood cells can become sickle shaped during mild exercise, making a person breathless (b) In cystic fibrosis, mucus can build up in the lungs, which requires inhalers to reduce the symptoms (c) In haemophilia, blood fails to clot, making minor problems such as a bleeding nose much more serious*

> 🔍 **Hint**
>
> Remember **ROLF** – mutations are of **R**andom **O**ccurrence and of **L**ow **F**requency.

> 🔬 **Make the link**
>
> There is more about cancer in Chapter 1.1 on page 17.

> 🔬 **Make the link**
>
> There is more about somatic cells in Chapter 1.1 on page 12.

(a) **(b)**

Figure 1.4.1 *We are widely encouraged (a) to use sunblock and (b) avoid tobacco*

> 🔬 **Make the link**
>
> There is more about the inheritance of genetic conditions in Chapter 2.4 on page 131.

Hint

Remember **SID** – **S**ubstitution, **I**nsertion and nucleotide **D**eletion – the single gene mutations.

Substitution

A single gene mutation in which one nucleotide is replaced by another.

Single gene mutations

Single gene mutations involve the alteration of a DNA nucleotide sequence as a result of the processes called substitution, insertion or deletion of nucleotides.

Nucleotide substitution

Nucleotide **substitutions** involve one nucleotide in a gene sequence being replaced by another incorrect one as shown in **Figure 1.4.3**. This means that one DNA codon is changed to another. This may have various different effects on the protein which is then expressed, depending on which codon is affected.

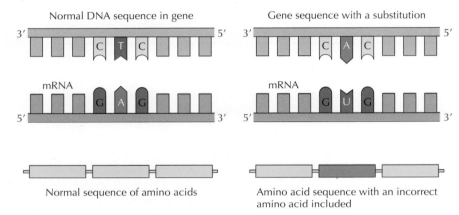

Figure 1.4.3 *The effects of a single nucleotide substitution in the gene encoding haemoglobin*

There are three types of substitution mutations:

- **Missense:** If the substitution affects one of the amino acid-coding codons in the sequence, then an incorrect amino acid may be placed into the protein at one place. This minor effect on the structure of the protein could have a major impact on the functioning of the protein. In sickle-cell disease, the gene which encodes the protein haemoglobin is affected by a missense mutation.

- **Nonsense:** If the substitution affects a codon by changing it to a stop codon, the premature stop codon may result in a shorter protein and a serious impact on its function. This is the cause of Duchenne muscular dystrophy.

- **Splice-site mutation:** If the substitution affects a codon at the splice site between an intron and an exon, this may result in the intron being retained or the exon not being included in the mature transcript. This is the cause of beta thalassemia.

Nucleotide insertion and deletion

In nucleotide **insertion** or **deletion**, there is a change in the number of nucleotides in the sequence, resulting in a change in the reading frame and causing all the codons after the mutation to be affected, as shown in **Figure 1.4.4**. These effects are known as frameshift mutations. Frameshift mutations have a major effect on the structure of the protein produced since all the amino acids after the mutation are likely to be affected.

> 📖 **Nucleotide insertion**
>
> A single gene mutation in which an additional nucleotide is placed into a DNA sequence.

Normal mRNA

Reading frames

AUG AAG UUU GGC UAA
Normal

Deletion of one U nucleotide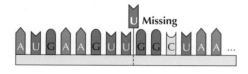

AUG AAG UUG GCU AA...
Frameshifted

Insertion of an extra U nucleotide

AUG UAA GUU UGG CUA
Frameshifted

Figure 1.4.4 *Effects of deletion and insertion of nucleotides on a gene sequence and the reading frames they produce – all the codons after the mutation are affected*

Chromosome structure mutation

Chromosome structure mutations involve significant changes to chromosomes. These mutations often arise during cell division when chromosomes are moving around in a dividing cell.

- **Deletion:** A piece of a chromosome becomes detached and lost. A deletion from chromosome 5 causes Cri-du-chat syndrome.

- **Duplication:** A piece of one chromosome becomes part of its **homologous** partner, resulting in a chromosome with two copies of a sequence.

- **Inversion:** A piece of chromosome is turned through 180 degrees. An inversion with a gene for a clotting factor is one cause of haemophilia.

- **Translocation:** A piece of one chromosome is detached and then joins onto another non-homologous chromosome in the genome. A translocation between two chromosomes is the cause of chronic myeloid leukaemia.

> 📖 **Nucleotide deletion**
>
> A single gene mutation involving the removal of a nucleotide from a sequence of DNA.

> **Make the link**
>
> There is more about cell division and chromosome movements in Chapter 1.1 on page 12.

> 📖 **Homologous chromosomes**
>
> Two chromosomes which carry the same sequence of genes.

The changes resulting from chromosome structure mutations are shown in **Figure 1.4.5**. These substantial changes often make the mutations lethal.

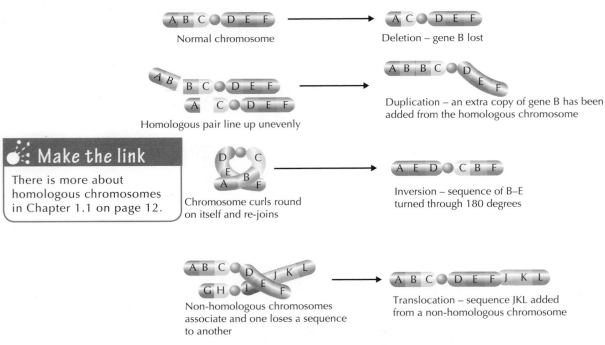

Make the link

There is more about homologous chromosomes in Chapter 1.1 on page 12.

Figure 1.4.5 *Chromosome mutations arising as a result of errors occurring as chromosomes move around in cells*

GO! Activity 1.4.1 Work individually to ...

Structured questions

1. **Explain** what is meant by the term 'mutation'. 2

2. **Name** the **three** types of single gene mutation and describe how they each change the genes they affect. 3

3. **Explain** what is meant by the following categories of substitution mutation:

   a) Missense

   b) Nonsense

   c) Splice-site 3

4. Ultraviolet (UV) light is known to increase the rate of mutation in cells by damaging their DNA. UV-sensitive yeast cells can be used to show the frequency of mutation caused by UV light because cells of this yeast variety cannot repair DNA damage caused by the UV and the cells die.

   An investigation into the effects of a sunblock cream on the rates of mutation in UV-sensitive yeast cells was carried out. UV-sensitive yeast cells were grown on nutrient agar in Petri dishes and then exposed to UV light for different times as shown in the diagram.

(*continued*)

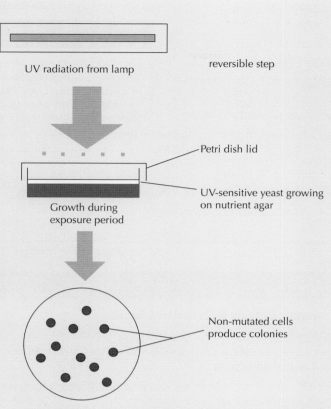

UV radiation from lamp

reversible step

Petri dish lid

UV-sensitive yeast growing on nutrient agar

Growth during exposure period

Non-mutated cells produce colonies

The procedure was repeated but the dishes were coated in a sunblock before exposure. The results are shown in the table.

Length of UV exposure time (minutes)	Number of yeast colonies growing in the dishes	
	Control (no sunblock)	Dishes coated with sunblock
0	120	120
10	64	120
20	32	108
30	16	84
40	4	76
50	2	68
60	0	64

a) **Give two** variables which should be kept the same for the control dishes. 2

b) **Give one** additional variable which should be kept constant for the dishes coated with sunblock. 1

c) On a separate piece of graph paper, **plot a line graph** to show all the results of this investigation. 3

d) **Calculate** the percentage increase in number of colonies after 30 minutes of exposure to UV light when sunblock was used to coat the dish. 1

e) It was concluded that the sunblock was an effective protection from UV light.

   Give evidence that this conclusion is **not** true? 1

f) **Describe** how the reliability of this investigation could be improved. 1

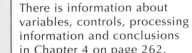

Make the link

There is information about variables, controls, processing information and conclusions in Chapter 4 on page 262.

Extended response questions

1. **Describe** the effects of frameshift mutations on the proteins they encode. 4
2. **Give an account** of the different types of chromosome mutation. 8

 Activity 1.4.2 Work in pairs to ...

1. **Practical activity: Effect of UV exposure time on mutation rate in yeast**

   You will need: six Petri dishes with nutrient agar, a 10^{-4} dilution of UV-sensitive yeast suspension, a glass spreader, tape, a marker pen, a UV lamp, a Bunsen and mat, a disposable 1 ml sterile pipette, aluminium foil and a stopwatch.

   ⚠ Your teacher will give specific safety instructions for carrying out this experiment and demonstrate the aseptic techniques needed.

   • Wash your hands with soap and water before and after doing the experiment.
   • Ensure that your bench area is wiped down with disinfectant before and after doing the experiment.
   • Ensure that all used apparatus and material is disposed of as instructed by your teacher.

   Method

   • Using the aseptic technique and a disposable sterile pipette, transfer 100 μl of yeast culture into each Petri dish.
   • Using the aseptic technique, spread the culture evenly over the agar surface in each dish using a glass spreader.
   • Tape each dish closed with two small pieces of tape.
   • Label each plate with your initials and its UV exposure time (0, 10, 20, 30, 40 and 50 minutes).
   • Wrap the 0 minutes plate in aluminium foil immediately it is set up.
   • Irradiate each plate for the appropriate time under a UV lamp.
   • Wrap the plates in aluminium foil as they are removed from the UV lamp.
   • Incubate all five plates for 2–3 days at 30°C.
   • Count the number of distinct colonies on each plate and record the numbers in a table.
   • On a separate sheet of graph paper, draw a line graph to show how the UV exposure time affected the mutation rate of yeast.

(continued)

A set of sample results is given below, to use if you are unable to carry out the experiment or if it did not seem to work properly or give the expected results.

UV light exposure time (minutes)	Mutation rate (number of colonies visible after exposure and incubation)
0	110
10	40
20	20
30	10
40	0
50	0

Answer the following questions

a) **Identify** the independent variable in this experiment.   1

b) **Describe** how the class results could be treated to make the results obtained more reliable.   1

c) **Give** a conclusion which can be drawn from the experimental results.   1

Assignment Support

You could use this experimental technique to generate data for your assignment:

- You could investigate many aspects of yeast exposure to UV light, such as the effects of different sunblocks or thickness of sunblock layers.

- You could use a copy of the grid on page 288 to plan an assignment based on this experiment.

2. Flashcard activity

You will each need: a set of blank flashcards (A7 cards) and a stopwatch.

- Find the glossary terms for this chapter – they are the **black** typeface and **red** typeface terms. Using your blank cards, you should each make a set of flashcards for these terms – write the term on one side and the definition on the other. You will find the definitions in the chapter.

- Shuffle your cards and lay them out in a column, some showing terms and some showing definitions – you decide. Your partner should match their cards with yours, laying their cards in a column beside yours to give the corresponding term or definition. Time how long they take to do this.

- Now swap roles – your partner should lay out their cards and you should try the matching exercise while your partner times you.

- You should each keep your set of flashcards as a revision tool for later.

Activity 1.4.3 Work as a group to ...

1. **Design and make a chromosome mutation relief poster.**

   You will need: a sheet of A3 card, a supply of two different coloured pipe-cleaners, scissors, glue stick or sticky tape and marker pens.

   Work as a group to design and make a relief poster to illustrate the different types of chromosome mutation – you could divide the mutations among the group and do one each.

   Ensure that each mutation is illustrated by a pipe-cleaner model of what happens to the DNA sequences.

   On the poster, give examples of human conditions caused by the different types of chromosome mutation.

Learning checklist

After working on this chapter, I can:

Knowledge and understanding

1. State that mutations are changes in the DNA that can result in no protein or an altered protein being synthesised.

2. State that single gene mutations involve the alteration of a DNA nucleotide sequence as a result of the substitution, insertion or deletion of nucleotides.

3. State that nucleotide substitutions include missense, nonsense and splice-site mutations.

4. State that missense mutations result in one amino acid being changed for another. This may result in a non-functional protein or have little effect on the protein.

5. State that nonsense mutations result in a premature stop codon being produced, which results in a shorter protein.

6. State that splice-site mutations result in some introns being retained or some exons not being included in the mature transcript.

7. State that nucleotide insertions or deletions result in frameshift mutations.

8. State that frameshift mutations cause all of the codons and all of the amino acids after the mutation to be changed. This has a major effect on the structure of the protein produced.

9. State that chromosome structure mutations include duplication, deletion, inversion and translocation.

10. State that duplication is where a section of a chromosome is repeated with sequences from its homologous partner.

11. State that deletion is where a section of a chromosome is removed.

12. State that inversion is where a section of a chromosome is reversed.

13. State that translocation is where a section of a chromosome is added to a chromosome, not its homologous partner.

14. State that the substantial changes in chromosome mutations often make them lethal.

Skills

1. *Plan experimental work through identification of variables.*

2. *Present information as a compound line graph.*

3. *Process information through calculation of a percentage increase.*

4. *Draw a conclusion from experimental results.*

5. *Evaluate supporting evidence to decide if a conclusion is valid.*

6. *Evaluate the reliability of an experiment.*

1.5 Human genomics

You should already know:

- The base sequence of DNA determines amino acid sequence in proteins.
- A gene is a section of DNA which codes for a protein.

Learning intentions

- Explain what makes up the genome of an organism.
- Describe methods used to analyse a genome.
- Explain how an individual's genome can be used in medicine.

The human genome

The **genome** of an organism is its entire hereditary information encoded in its DNA. A genome is made up of **genes** and other DNA sequences. Genes are DNA sequences which code for proteins and the other DNA sequences in the genome are said to be non-coding as they do not encode proteins. Non-coding sequences include introns and sequences which are transcribed to tRNA and rRNA. The human genome is carried in the chromosomes, which are found in the nucleus of cells as shown in **Figure 1.5.1(a)**. The genomic sequences in DNA are found strung out along the chromosomes as shown in **Figure 1.5.1(b)**.

(a)

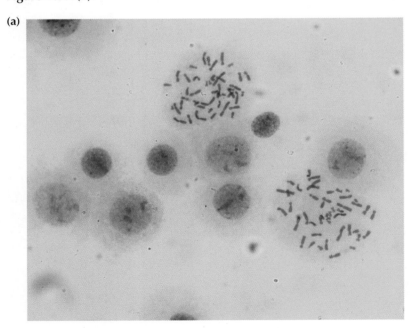

(b)

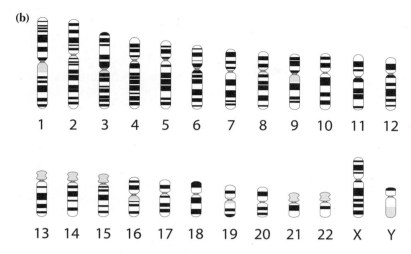

Figure 1.5.1 *(a) Photograph of stained human chromosomes in cells under a microscope (b) Drawings of each of the human chromosomes arranged in order and showing the banding patterns which have been identified by the mapping of the genome*

The Human Genome Project

The human genome has been the subject of intense investigation over recent times. The Human Genome Project (HGP) aimed to give a complete and accurate sequence of the three billion DNA base pairs that make up the human genome and to locate all of the estimated 20,000 to 25,000 human genes which are scattered through it. The project also aimed to develop new tools to locate and analyse genomic data and to make this information widely available. Advances in genetics have consequences for individuals and society and the project was committed to exploring the consequences of genomic research through its Ethical, Legal, and Social Implications (ELSI) programme. The project also aimed to sequence the genomes of several other species that are important to medical research, such as the mouse and the fruit fly.

The project began in 1990 and published a finished version of the human genome sequence in 2004.

Genomic sequencing

Genomic sequencing allows the sequence of nucleotide bases to be determined for individual genes and entire genomes. Computer programs can be used to identify base sequences by looking for sequences similar to known genes. Computer and statistical analyses known as **bioinformatics** are required to compare sequence data.

Figure 1.5.2 *Bioinformatics uses statistics and computations to analyse and compare sequence data*

📖 Pharmacogenetics

The study of how individuals respond to specific drugs based on their individual genome.

Figure 1.5.3 *The future of personalised medicine – will doctors routinely use the genomes of their patients to make decisions about treatments and drugs?*

Make the link

There is more about proteins in Chapter 1.3 on page 39.

Personal genomics

An individual personal genome can be analysed to predict the likelihood of developing certain diseases. **Pharmacogenetics** uses personal genome information to select the most effective drugs and dosages to treat disease in the individual and allows the development of **personalised medicine**.

Personalised medicine

Personalised medicine involves tailoring decisions about interventions, treatments and drugs, based on the predicted response by the patient. An individual's genome could provide evidence to guide these decisions. **Figure 1.5.4** shows how a group of different individual patients may react differently to the same prescribed drug. If their genomes were the basis of the differences, drug choices could be better informed.

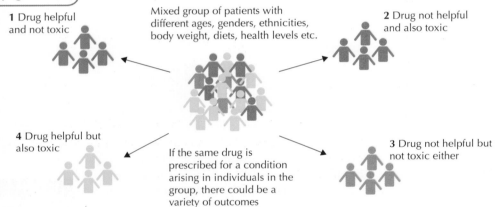

Figure 1.5.4 *Possible reactions of a group of individuals to the same prescribed drug*

Activity 1.5.1 Work individually to ...

Structured questions

1. The human genome consists of genes and other DNA sequences.
   a) **State** what is meant by a gene.   1
   b) **Give** an example of another DNA sequence.   1
2. **Give** the meaning of the following terms:
   a) Bioinformatics   1
   b) Pharmacogenetics   1
3. The table gives information about the estimated genome size, chromosome number and number of genes of different species, including humans.

(continued)

Species	Estimated genome size (millions of base pairs)	Chromosome number	Estimated number of genes in genome
Human (Homo sapiens)	3,000	46	25,000
Laboratory mouse (Mus musculus)	2,600	40	25,000
Thale cress (Arabidopsis thaliana)	100	10	35,000
Roundworm (Caenorhabditis elegans)	97	12	19,000
Fruit fly (Drosophila melanogaster)	137	8	13,000
Yeast (Saccharomyces cerevisiae)	12·1	32	6,000
Bacterium (Escherichia coli)	4·6	1	3,200

a) It was concluded that the species with the largest genomes had the biggest number of genes.

Describe evidence from the table which does **not** support this conclusion. 1

b) Calculate how many times bigger the human genome is compared with that of the bacterium Escherichia coli. 1

c) i. Assuming the entire genome of humans was made up of genes, calculate the average number of base pairs per gene. 1

ii. Explain why it is wrong to assume that a genome is made up only of genes. 1

d) i. Assuming that there are the same number of genes on each chromosome, calculate the number of genes on each human chromosome. 1

ii. Using information in Figure 1.5.1(b), explain why it would be wrong to assume that there are the same number of genes on each human chromosome. 1

Make the link

There is more information about processing information in calculations and drawing conclusions in Chapter 4 on page 263.

Extended response question

1. Give an account of personalised medicine. 4

Activity 1.5.2 Work in pairs to …

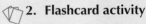

1. Research sequencing the human genome.

Visit the following web page:

www.genome.gov/human-genome-project/Completion-FAQ

Select the question 'What is DNA sequencing?' and read the answer given.

a) Work together to construct a flowchart to show the steps taken in the BAC method for sequencing the human genome.

b) Find out whose genome was sequenced in the project and who owns the human genome.

2. Flashcard activity

You will each need: a set of blank flashcards (A7 cards) and a stopwatch.

- Find the glossary terms for this chapter – they are the **black** typeface and **red** typeface terms. Using your blank cards, you should each make a set of flashcards for these terms – write the term on one side and the definition on the other. You will find the definitions in the chapter.

- Shuffle your cards and lay them out in a column, some showing terms and some showing definitions – you decide. Your partner should match their cards with yours, laying their cards in a column beside yours to give the corresponding term or definition. Time how long they take to do this.

- Now swap roles – your partner should lay out their cards and you should try the matching exercise while your partner times you.

- You should each keep your set of flashcards as a revision tool for later.

Activity 1.5.3 Work as a group to …

1. Design and make a poster on the applications of bioinformatics.

You will need: internet access, a piece of A3 card and coloured marker pens.

Use online sources to identify five applications of bioinformatics. For each application, write a sentence to describe what it is.

Make up a poster listing your five applications and explaining what they are.

Learning checklist

After working on this chapter, I can:

Knowledge and understanding

1. State that the genome of an organism is the genetic information that is encoded into its DNA and can be inherited by its offspring.

2. State that an organism's genome is made up of genes, which are DNA sequences that code for proteins and other DNA sequences that do not code for proteins.

3. State that non-coding sequences include those that are transcribed into RNA but are not translated.

4. State that tRNA and rRNA are non-translated forms of RNA.

5. State that in genomic sequencing the sequence of nucleotide bases can be determined for individual genes and entire genomes.

6. State that computer programs can be used to identify base sequences by looking for sequences similar to known genes.

7. State that computer and statistical analyses (bioinformatics) are required to compare sequence data.

8. State that an individual's genome can be analysed to predict the likelihood of developing certain diseases.

9. State that pharmacogenetics is the use of genome information in the choice of drugs.

10. State that personalised medicine is based on an individual's genome.

11. State that an individual's personal genome sequence can be used to select the most effective drugs and dosage to treat their condition.

Skills

1. *Evaluate a conclusion using evidence from tabulated data.*

2. *Process information by calculating averages from tabulated data.*

1.6 Metabolic pathways

You should already know:

- Enzymes function as biological catalysts and are made by all living cells.
- Enzymes speed up cellular reactions and are unchanged in the process.
- The shape of the active site of an enzyme molecule is complementary to its specific substrate(s).
- Enzyme action results in product(s).
- Enzymes can be involved in degradation and synthesis reactions.
- Each enzyme is most active in its optimum conditions.
- Enzyme activity can be affected by temperature and pH.
- Enzymes can be denatured, resulting in a change in the shape of their active site, which will affect the rate of reaction.
- A respirometer can be used to measure the rate of respiration.

Learning intentions

- Describe the role of metabolic pathways in a cell.
- Describe metabolic pathways in terms of integrated and controlled pathways of enzyme-catalysed reactions within a cell which can have reversible steps, irreversible steps and alternative routes.
- Describe the differences between an anabolic pathway and a catabolic pathway.
- Describe the different ways in which metabolic pathways can be controlled.
- Explain the induced fit model of enzyme action.
- Describe the role of enzymes in metabolic pathways.
- Describe the effects of substrate and product concentration on the direction and rate of enzyme reactions.
- Explain the control of metabolic pathways through competitive, non-competitive and feedback inhibition of enzymes.

- Be familiar with the use of substrate concentration or inhibitor concentration to alter the rate of an enzyme reaction.

- Describe the use of simple respirometers and gas probes to measure metabolic rate.

Introducing metabolic pathways

All the chemical reactions that take place in the cells of a living organism are collectively called **metabolism**.

A **metabolic pathway** is a series of connected and integrated chemical reactions, each controlled by different enzymes which act as biological catalysts as shown in **Figure 1.6.1**.

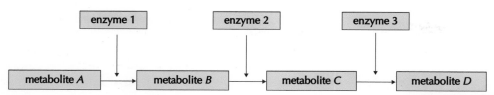

Figure 1.6.1 *A metabolic pathway involving metabolites A–D and enzymes 1–3*

Make the link

There is more about polypeptides and protein synthesis in Chapter 1.3 on page 35.

Metabolic pathways can have reversible steps, irreversible steps and alternative routes as shown in **Figure 1.6.2**. Reactions within metabolic pathways can be anabolic or catabolic.

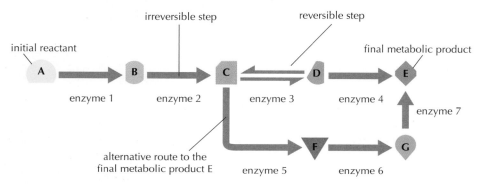

Figure 1.6.2 *Metabolic pathway with one reversible step, several irreversible steps and an alternative route to the final metabolic product*

Anabolic reactions

Anabolic reactions build up large molecules from small molecules and require energy in the form of ATP. The synthesis of a polypeptide from amino acids is an example of an anabolic reaction.

The general characteristics of an anabolic reaction are shown in **Figure 1.6.3**.

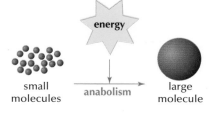

Figure 1.6.3 *Summary of the features of an anabolic reaction*

Catabolic reactions

Catabolic reactions break down large molecules into smaller molecules and release energy. Aerobic respiration is an example of a catabolic reaction which involves the breakdown of glucose and the release of energy. In this example, most of the energy released is used to produce ATP, which acts as a store of chemical energy, although some energy is also released as heat.

Make the link

There is more about aerobic respiration and ATP production in Chapter 1.7 on page 74.

A summary of a catabolic reaction showing breakdown of a larger molecule into smaller molecules and the energy released is shown in **Figure 1.6.4**.

Catabolic and anabolic pathways are linked through chemical energy. The chemical energy in ATP released from catabolic pathways can be used in anabolic pathways to synthesise the new substances required by the cell as shown in **Figure 1.6.5**.

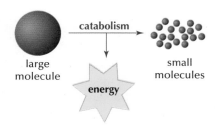

Figure 1.6.4 *Summary of the features of a catabolic reaction*

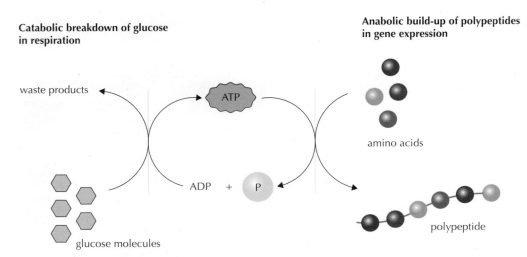

**Catabolic breakdown of glucose
in respiration**

**Anabolic build-up of polypeptides
in gene expression**

Figure 1.6.5 *Catabolic and anabolic pathways linking through ATP to synthesise new substances – here glucose is broken down catabolically to release energy in ATP for the anabolic synthesis of a polypeptide*

Control of metabolic pathways

Metabolic pathways are controlled by the presence or absence of particular enzymes and the regulation of the rate of reaction of key enzymes. Metabolic pathways may be controlled by the production, activation or inhibition of particular enzymes.

Since metabolic pathways are a series of enzyme-catalysed reactions, it is possible to control these reactions by regulating the activity of the enzymes involved. If the enzyme is present, the reaction proceeds but if the enzyme is absent, the reaction stops.

Like all proteins, an enzyme is coded for by a gene which may be 'switched on' or 'switched off' depending on whether or not the enzyme is required. For example, the gene that codes for an enzyme involved in digestion may not be switched on if the substrate is absent because the enzyme is not required. The vital process of cellular respiration goes on all the time and so the genes associated with the many enzymes of cellular respiration are constantly switched on.

Enzyme action

📖 **Induced fit**

Changes to an enzyme's active site brought about by its substrate.

📖 **Affinity**

The attraction between the active site of an enzyme and its substrate.

Enzymes only act on one specific substrate which is complementary to and fits into the active site. The **active site** is the location on the enzyme where the substrate molecule binds and the chemical reaction takes place. This action is called an **induced fit** and makes use of the fact that the active site is not a rigid arrangement but can change shape. Induced fit occurs when the active site changes shape to fit the substrate more closely after the substrate binds.

The substrate molecule(s) have a high **affinity** for the active site and the subsequent products have a low affinity, allowing them to leave the active site as shown in **Figure 1.6.6**.

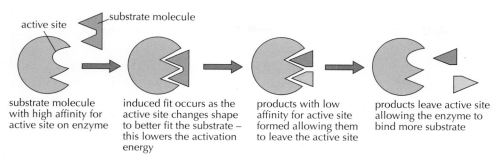

active site

substrate molecule

substrate molecule with high affinity for active site on enzyme

induced fit occurs as the active site changes shape to better fit the substrate – this lowers the activation energy

products with low affinity for active site formed allowing them to leave the active site

products leave active site allowing the enzyme to bind more substrate

Figure 1.6.6 *Enzyme affinity and induced fit mechanism of a catabolic reaction*

Activation energy

In order to make some reactions take place, a small input of energy is required to activate the process. The energy input required to make the chemical reaction take place is called the **activation energy**. When the reactants have reached the necessary activation energy, the reaction can proceed and products are formed.

In an enzyme-catalysed reaction, the presence of the enzyme lowers this activation energy, making it easier to start a reaction which otherwise might proceed slowly or not at all. This is shown in **Figure 1.6.7**. Notice that the enzyme does not affect the initial or final energy values for the reactants or products, nor does it affect the overall energy level in the products.

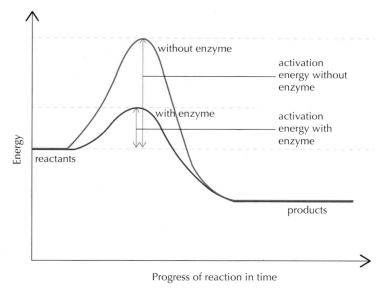

Figure 1.6.7 *The reduction of the activation energy due to the presence of an enzyme*

Factors affecting enzyme action

The direction and rate of an enzyme-catalysed reaction are influenced by both the concentration of the substrate and the concentration of the product(s). **Figure 1.6.8** shows the effect of increasing the substrate concentration on the rate of an enzyme-catalysed reaction when other variables are kept constant.

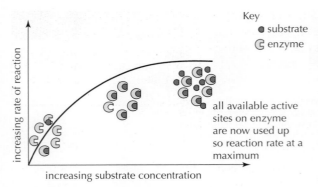

Figure 1.6.8 *The effect of substrate concentration on the rate of an enzyme-controlled reaction*

As the substrate concentration increases, it drives the reaction in the direction of the end product and increases the rate of the reaction. More and more of the enzyme's active sites become used, until a point is reached when no more active sites are available. The rate of reaction is at maximum and now remains constant in spite of further availability of substrate. At this point, some other factor needs to be altered to make the reaction go any faster.

Feedback inhibition

Another way that an enzyme-catalysed reaction can be regulated occurs when the product of the last reaction in a metabolic pathway inhibits the enzyme which catalyses the first reaction of the pathway. This is called **end-product** or **feedback inhibition** and it ensures that levels of products meet the supply requirements of the cell without excess being produced, as shown in **Figure 1.6.9**.

> **Hint**
>
> Remember that other factors, such as pH and temperature, can also affect enzyme activity.

> **Feedback/ End-product inhibition**
>
> Enzyme inhibition caused by the presence of an end product of a metabolic pathway.

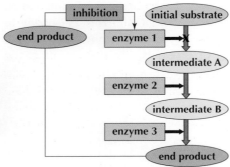

Figure 1.6.9 *Feedback/End-product inhibition of a metabolic pathway*

The product of the last reaction binds to a site other than the active site of the enzyme which catalyses the first reaction, changing the shape of the active site as shown in **Figure 1.6.10**.

This means that the attachment of the first substrate to the first enzyme is now less likely and the reaction slows down or stops altogether. However, when the supply of the final product falls below a threshold level, its attachment to the first enzyme no longer occurs and the active site can once again bind with its normal substrate to allow the reaction to proceed.

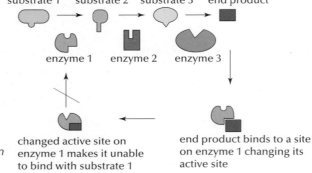

Figure 1.6.10 *End-product inhibition changes the shape of the active site of an earlier enzyme*

changed active site on enzyme 1 makes it unable to bind with substrate 1

end product binds to a site on enzyme 1 changing its active site

The advantage of using end-product inhibition is to allow control of a metabolic pathway. If end products are building up, the cell can save energy by stopping making more, but if the end product is absent or scarce the cell can re-start synthesis.

Competitive and non-competitive inhibition

The activity of many enzyme reactions can be controlled by the binding of specific molecules called **inhibitors**. Inhibition of an enzyme is a way of controlling biological systems by affecting the enzyme-catalysed reactions associated with them. There are two important types of inhibitors – competitive and non-competitive inhibitors.

Competitive inhibitors

Molecules which compete with the normal substrate for the active site of an enzyme but which don't permanently disable the enzyme are called **competitive inhibitors**. Competitive inhibitors are very similar in their shape to the normal substrate of a particular enzyme and will compete with these substrate molecules to occupy the active site as shown in **Figure 1.6.11**.

The enzyme either binds to the normal substrate or the competitive inhibitor, but not both. Since a competitive inhibitor competes with the normal substrate for the active site, increasing the substrate concentration in the presence of a competitive inhibitor will cause the rate of enzyme activity to increase again as shown in **Figure 1.6.12**.

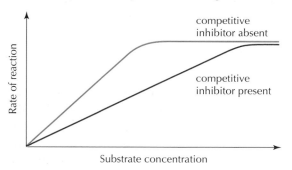

Figure 1.6.12 *The effect of increasing substrate concentration in the presence of a competitive inhibitor*

Non-competitive inhibitors

Some enzymes have two sites: one for the normal substrate and another for an inhibitor. When the inhibitor binds to the enzyme it causes the shape of the active site to be altered so that the enzyme will no longer fit the substrate, as shown in **Figure 1.6.13**. Such inhibitors are termed **non-competitive** because they do not compete for the active site. Instead they bind to a site somewhere other than the active site but by doing so the enzyme can no longer bind to the normal substrate.

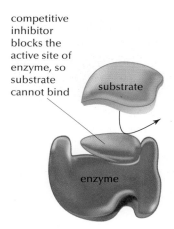

competitive inhibitor blocks the active site of enzyme, so substrate cannot bind

substrate

enzyme

Figure 1.6.11 *The action of a competitive inhibitor occupying the active site of an enzyme*

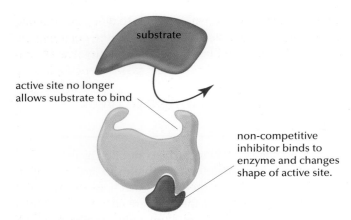

Figure 1.6.13 *The action of a non-competitive inhibitor changes the shape of the active site*

Since the inhibitor in this case alters the shape of the active site, increasing the substrate concentration will not increase the rate of the enzyme-catalysed reaction, as shown in **Figure 1.6.14**. This type of inhibition is irreversible.

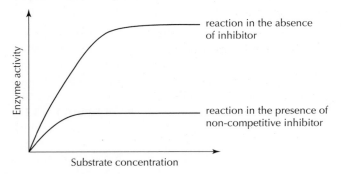

Figure 1.6.14 *The effect of increasing substrate concentration in the presence of a non-competitive inhibitor*

Figure 1.6.15 shows the effect of increasing the concentration of a competitor inhibitor on the rate of reaction of an enzyme.

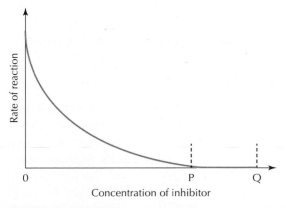

Figure 1.6.15 *The effect of competitive inhibitor concentration on the rate of an enzyme-controlled reaction*

As the concentration of the competitive inhibitor is increased, the rate of reaction decreases because more active sites are occupied by inhibitor molecules. Between points P and Q, in **Figure 1.6.15**, the reaction has stopped because all the active sites are occupied by inhibitors.

Metabolic rate

 A living organism constantly uses energy to survive and the rate at which this energy is used is called **metabolic rate**. The energy to drive metabolism comes from the breakdown of glucose during cell respiration. Since oxygen is used up and carbon dioxide and energy are released, it is possible to measure the metabolic rate of a cell by measuring how much:

- oxygen is used up in a given period of time
- carbon dioxide is produced in a given period of time
- energy (in the form of heat) is released in a given period of time

Measuring metabolic rate is one of the techniques you need to be familiar with for your exam.

 A **respirometer** usually measures the consumption of oxygen over a period of time, as shown in **Figure 1.6.16**.

📖 Metabolic rate

The amount of energy used by an organism in a given period of time.

📖 Respirometer

Apparatus that measures oxygen uptake as an indirect measurement of metabolic rate.

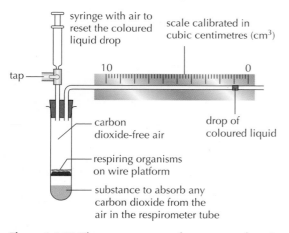

Figure 1.6.16 *The measurement of oxygen uptake using a simple respirometer – the respiring organisms use oxygen from the air, drawing the drop of coloured liquid along the scale; any carbon dioxide they produce is absorbed by the substance so does not affect the movement of the coloured liquid*

Figure 1.6.17 shows how oxygen or carbon dioxide probes can give information on the changes to gas levels caused by respiring material in a closed container.

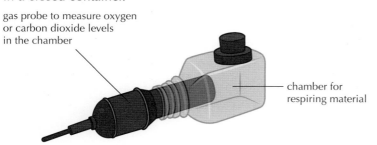

Figure 1.6.17 *Using a digital gas probe to measure changes in oxygen or carbon dioxide levels*

A **calorimeter** is a device that can directly measure the heat produced by an organism over a given period of time. A calorimeter with a human subject is shown in **Figure 1.6.18**.

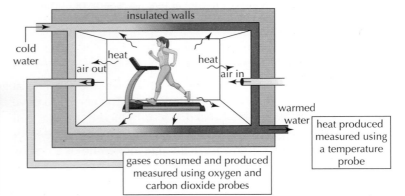

📖 Calorimeter

A device that gives a direct measurement of metabolic rate by measuring the heat produced in a given period of time.

Figure 1.6.18 *A calorimeter designed to measure metabolic rate in a human subject*

GO! Activity 1.6.1 Work individually to ...

Structured questions

1. Metabolic pathways are integrated and controlled pathways of anabolic and catabolic reactions.
   a) **Describe** what is meant by a metabolic pathway.   1
   b) **Describe** the features of an anabolic reaction.   1
   c) **Describe** the features of a catabolic reaction.   1

2. **Explain** the induced fit model of enzyme action.   2

3. The direction and rate of an enzyme-catalysed reaction are influenced by both the concentration of the substrate and the concentration of the product(s).

   Describe the effect of an increase in product concentration on the direction and rate of an enzyme reaction.   2

4. The activity of many enzymes can be controlled by the binding of specific molecules called inhibitors.
   a) **Describe** the action of a competitive inhibitor.   1
   b) **Describe** the action of a non-competitive inhibitor.   1
   c) **Explain** feedback control of a metabolic pathway.   2

5. The diagram shows an enzyme-catalysed reaction taking place in the presence of an inhibitor.
   a) **Identify** the letter that represents the substrate molecule.   1
   b) **Explain** the type of inhibition represented in this reaction.   3

6. An investigation was carried out into the effect of a competitive inhibitor on the activity of phosphatase at different substrate concentrations. Phosphatase is an enzyme which catalyses the degradation of phenolphthalein phosphate as shown.

   phenolphthalein phosphate → phenolphthalein + phosphate

         substrate   *products*

(continued)

Six test tubes each containing a different concentration of substrate were set up. The inhibitor and then the enzyme were added to each tube. The diagram shows the contents of each tube.

After 30 minutes, 1 cm³ of alkali was added to each tube. Phenolphthalein turns pink in the presence of alkali. The more phenolphthalein produced in the reaction, the more intense the pink colour and the higher the absorbance reading measured by a colorimeter.

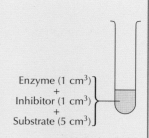

Enzyme (1 cm³)
+
Inhibitor (1 cm³)
+
Substrate (5 cm³)

The table below shows the results of the investigation.

Concentration of substrate (M)	Absorbance (units)
0·05	0·20
0·10	0·30
0·20	0·48
0·40	0·64
0·60	0·78
0·80	0·90

a) **Name** the independent variable in this investigation.   1

b) **Suggest** why alkali was not added to each tube at the start of the investigation.   1

c) **State two** other variables, not mentioned, which should be controlled to improve the validity of this investigation.   2

d) On a piece of graph paper, **draw a line graph** to show the data in the table.   2

e) It was concluded that increasing substrate concentration reduces the effect of the competitive inhibitor.

   Explain how the results of this investigation support this conclusion.   2

f) The inhibitor used in this investigation was a competitive inhibitor.

   Describe how competitive inhibition works.   1

g) **Suggest** how the results of this investigation would be different if a non-competitive inhibitor had been used.   1

Make the link

There is information about variables and drawing line graphs in Chapter 4 on page 263.

7. The graph below shows the mass of product resulting from an enzyme-controlled reaction.

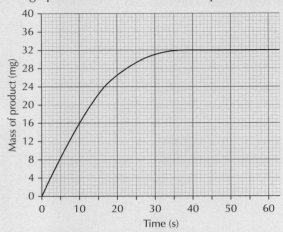

a) **State** how long it took for the mass of product produced to reach 50% of its final mass. 1

b) **Calculate** the rate of reaction over the first 10 seconds. 1

Extended response questions

1. **Give an account** of anabolic and catabolic reactions. 4

2. **Give an account** of the induced fit model of enzyme action and the control of metabolic pathways through competitive and non-competitive inhibition of enzymes. 8

GO! Activity 1.6.2 Work in pairs to ...

1. **Practical activity: The effect of catalase concentration on reaction rate**

   You will need: yeast cell (or algae) samples: stock culture, 10^{-1}, 10^{-2}, 10^{-3}, 10^{-4} dilutions; five pipettes, 1% hydrogen peroxide, measuring cylinder, six universal bottles, forceps, stopwatch, thermometer (to check that temperature is constant), filter paper discs and safety goggles.

   In this investigation you will use a culture of yeast cells in water. The culture has been diluted several times to produce a range of dilutions. The weakest dilution (10^{-4}) will contain the fewest cells and will therefore contain the least catalase. The strongest dilution (stock culture) will contain the most cells and will therefore contain the most catalase.

    Your teacher will give specific safety instructions for carrying out this experiment and demonstrate any techniques needed.

   • Wear eye protection while doing the experiment.

   • Wash your hands with soap and water before and after doing the experiment.

   • Ensure that all used apparatus and material is disposed of as instructed by your teacher.

   Method

   • Measure 20 cm³ of 1% hydrogen peroxide into a universal bottle.

   • Using a pipette, place one drop of yeast cell stock culture on to a filter paper disc.

   • Using forceps, drop the disc into the hydrogen peroxide solution.

   • Watching carefully from the side of the bottle, time (and note) how long it takes for the disc to rise to the surface.

(continued)

- Repeat the first four steps with the different yeast cell dilutions. Use a fresh filter paper disc, clean pipette and clean universal bottle each time.
- As a 'control', repeat the procedure using a drop of water instead of yeast cells on the filter paper disc.

Results and report

You should each write an account of what you did. Include a labelled diagram of the experiment. Use your results to construct a line graph and draw a conclusion about the effect of increasing the concentration of catalase on reaction rate.

Assignment Support

You could use this experimental technique to generate data for your assignment:

- You could investigate many aspects of the rates of enzyme-controlled reactions, such as the effects of temperature, pH, substrate concentration or inhibitors.
- You could use a copy of the grid on page 288 to plan an assignment based on this experiment.

Make the link

There is information about drawing line graphs in Chapter 4 on page 266.

2. Flashcard activity

You will each need: a set of blank flashcards (A7 cards) and a stopwatch.

- Find the glossary terms for this chapter – they are the **black** typeface and **red** typeface terms. Using your blank cards, you should each make a set of flashcards for these terms – write the term on one side and the definition on the other. You will find the definitions in the chapter.
- Shuffle your cards and lay them out in a column, some showing terms and some showing definitions – you decide. Your partner should match their cards with yours, laying their cards in a column beside yours to give the corresponding term or definition. Time how long they take to do this.
- Now swap roles – your partner should lay out their cards and you should try the matching exercise while your partner times you.
- You should each keep your set of flashcards as a revision tool for later.

Activity 1.6.3 Work as a group to …

1. Placemat activity: Inhibitors

You will need: a placemat template (Appendix 3) and four fine marker pens.

- Set the placemat in the middle of the table and each write your name in a section.
- Each participant should then write words that they think are related to *control of metabolic pathways through competitive, non-competitive and feedback inhibition* into their section of the placemat. Spend 2 minutes doing this.
- You should take it in turns to read out a word from your section. If everyone agrees it is related to the topic it can be copied into the centre section of the placemat. Continue until all words have been discussed.

- Working as a group, use all the words in the centre section to summarise your knowledge of control of metabolic pathways through competitive, non-competitive and feedback inhibition.

2. Ring of Fire: Metabolic pathways

*You will need: a printout of the **Ring of Fire: Metabolic pathways** question and answer card set and a stopwatch.*

- Your teacher will deal the question and answer cards until all the cards have been issued.
- Your teacher will nominate a student to read the first question aloud and will start the clock.
- The student with the correct answer card should read the answer and then ask their own question.
- This is repeated until all the questions are completed. The timer is stopped.
- You should repeat the whole game to try to improve your time.

Learning checklist

After working on this chapter, I can:

Knowledge and understanding

1. State that a metabolic pathway is a series of chemical reactions controlled by enzymes. ⬭ ⬭ ⬭

2. State that metabolic pathways are integrated and controlled pathways of enzyme-catalysed reactions within a cell. ⬭ ⬭ ⬭

3. State that metabolic pathways can have reversible steps, irreversible steps and alternative routes. ⬭ ⬭ ⬭

4. State that reactions within metabolic pathways can be anabolic or catabolic. ⬭ ⬭ ⬭

5. State that anabolic reactions build up large molecules from small molecules and require energy. ⬭ ⬭ ⬭

6. State that catabolic reactions break down large molecules into smaller molecules and release energy. ⬭ ⬭ ⬭

7. State that metabolic pathways are controlled by the presence or absence of particular enzymes and the regulation of the rate of reaction of key enzymes. ⬭ ⬭ ⬭

8. State that induced fit occurs when the active site of the enzyme changes shape to better fit the substrate after the substrate binds. ⬭ ⬭ ⬭

9. State that the substrate molecule(s) have a high affinity for the active site and the subsequent products have a low affinity, allowing them to leave the active site. ⬭ ⬭ ⬭

10. State that the energy required to initiate a chemical reaction is called the activation energy.

11. State that enzymes lower the activation energy required for chemical reactions.

12. State that the substrate and product concentration affect the direction and rate of enzyme reactions.

13. State that some metabolic reactions are reversible and the presence of a substrate or the removal of a product will drive a sequence of reactions in a particular direction.

14. State that competitive inhibitors bind at the active site of the enzyme, preventing the substrate from binding.

15. State that competitive inhibition can be reversed by increasing the substrate concentration.

16. State that non-competitive inhibitors bind away from the active site but change the shape of the active site, preventing the substrate from binding.

17. State that non-competitive inhibition cannot be reversed by increasing the substrate concentration.

18. State that feedback inhibition occurs when the end product in the metabolic pathway reaches a critical concentration.

19. State that the end product inhibits an earlier enzyme, blocking the pathway, and so prevents further synthesis of the end product.

20. Describe the use of simple respirometers to measure metabolic rates.

Skills

1. *Process information to calculate a rate of reaction.*

2. *Select information from a table and a graph.*

3. *Present data as a line graph.*

4. *Evaluate an experiment to identify variables that should be controlled.*

5. *Draw a conclusion from experimental data.*

1.7 Cellular respiration

You should already know:

- The chemical energy stored in glucose is released by all cells through a series of enzyme-controlled reactions called respiration.
- The energy released from the breakdown of glucose is used to generate ATP.
- The energy transferred by ATP can be used for cellular activities such as muscle cell contraction, cell division, protein synthesis and transmission of nerve impulses.
- Glucose is broken down to two molecules of pyruvate, releasing enough energy to yield two molecules of ATP.
- If oxygen is present, aerobic respiration takes place, and each molecule of pyruvate is broken down to carbon dioxide and water, releasing enough energy to yield a large number of ATP molecules.
- In the absence of oxygen, the fermentation pathway takes place. In animal cells, the pyruvate molecules are converted to lactate and in plant and yeast cells they are converted to carbon dioxide and ethanol.
- The breakdown of each glucose molecule via the fermentation pathway yields only the initial two molecules of ATP.
- Respiration begins in the cytoplasm. The process of fermentation is completed in the cytoplasm whereas aerobic respiration is completed in the mitochondria.

Learning intentions

- State the locations and describe the stages of glycolysis, the citric acid cycle and the electron transport chain in aerobic respiration.
- Describe the importance of the metabolic pathways of cellular respiration and the role of ATP in the transfer of energy.

Make the link

There is more about cell division, DNA replication and protein synthesis in Chapters 1.1, 1.2 and 1.3 on pages 12, 26 and 35.

Make the link

There is more about anabolism and catabolism in Chapter 1.6 on page 60.

Introducing cellular respiration

Cellular respiration is a catabolic pathway that takes place in the cells of living organisms to convert the chemical energy from glucose (and other substrates) into the high-energy molecule ATP (adenosine triphosphate).

The role of ATP in the transfer of energy

ATP provides cells with an instant usable source of energy. The role of ATP is to transfer chemical energy from cellular respiration to anabolic cellular processes which require energy, such as cell division, DNA replication and protein synthesis. As ATP makes this transfer, the ATP is broken down to adenosine diphosphate (ADP) and a molecule of inorganic phosphate (Pi) as shown in **Figure 1.7.1**.

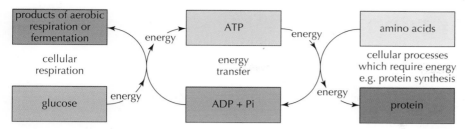

Figure 1.7.1 *Summary of the energy transfer system in cells*

The stages of aerobic respiration

Aerobic respiration involves three stages: glycolysis, the citric acid cycle and the electron transport chain, which take place in different locations in the cell.

Stage 1: Glycolysis

Glycolysis is the first stage of respiration and takes place in the cytoplasm of a cell. It is a multi-step metabolic pathway resulting in the breakdown of one molecule of glucose into two molecules of pyruvate. Each reaction in glycolysis is catalysed by its own enzyme. Glycolysis can take place with or without oxygen.

The first phase of glycolysis is called an **energy investment phase**. During this phase, two molecules of ATP are broken down to provide the phosphate groups needed for the **phosphorylation** of glucose and intermediates as shown in **Figure 1.7.2**.

The second phase of glycolysis is called the **energy pay-off phase**. During this phase, four molecules of ATP are generated. The generation of more ATP during the energy pay-off stage results in a net gain of two molecules of ATP as shown in **Figure 1.7.2**.

During the energy pay-off phase, **dehydrogenase** enzymes remove hydrogen ions (H^+) and electrons and binds them to the coenzyme **NAD** to form NADH. If oxygen is present, NADH transports the hydrogen ions (H^+) and electrons to the **electron transport chain** on the inner membrane of the mitochondria.

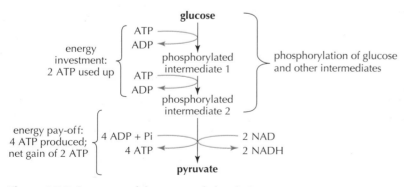

Figure 1.7.2 *Summary of the stages of glycolysis*

🔍 Hint

Glycolysis means the splitting of glucose.

📖 Energy investment phase

The first stage in glycolysis, which uses two molecules of ATP to phosphorylate glucose and intermediates.

📖 Phosphorylation

The addition of phosphate to a molecule.

📖 Energy pay-off phase

The second stage in glycolysis, which produces four molecules of ATP, giving a net gain of two molecules of ATP.

📖 Dehydrogenase

An enzyme which removes hydrogen ions (H^+) and electrons from their substrates and binds them to NAD.

📖 NAD

The coenzyme that transports hydrogen ions (H^+) and electrons to the electron transport chain on the inner membrane of the mitochondria.

Citric acid cycle

The second stage of aerobic respiration that takes place in the matrix of the mitochondria.

Acetyl group

Produced by the breakdown of pyruvate and joins with oxaloacetate to form citrate in the citric acid cycle.

Oxaloacetate

Substance which combines with the acetyl group in the citric acid cycle to form citrate.

Electron transport chain

A chain of proteins embedded in the inner membranes of mitochondria that transport electrons.

ATP synthase

The enzyme embedded in the inner membrane of the mitochondria that synthesises ATP as hydrogen ions (H$^+$) flow through it.

Oxygen

Acts as the final electron acceptor in aerobic respiration and joins with hydrogen ions (H$^+$) to form water.

Stage 2: Citric acid cycle

The **citric acid cycle** occurs in the matrix of the mitochondria. In aerobic conditions, pyruvate enters the central cavity of the mitochondria. It is then broken down to an **acetyl group** that combines with coenzyme A to form acetyl coenzyme A. In the citric acid cycle the acetyl group from acetyl coenzyme A combines with **oxaloacetate** to form citrate. Citrate is gradually converted back into oxaloacetate during a series of enzyme-controlled steps, which results in the generation of one molecule of ATP and the release of carbon dioxide as shown in **Figure 1.7.3**.

Dehydrogenase enzymes remove hydrogen ions (H$^+$) and electrons and pass them to the coenzyme NAD, forming NADH. The hydrogen ions (H$^+$) and electrons from NADH are passed to the electron transport chain on the inner membrane of the mitochondria.

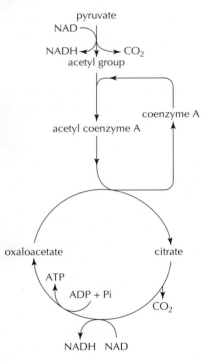

Figure 1.7.3 *Progress of pyruvate under aerobic conditions in the matrix of the mitochondria*

Stage 3: Electron transport chain

The electron transport chain takes place in the mitochondria of the cell. It is a series of carrier proteins attached to the inner mitochondrial membrane. **Electrons** are passed along the electron transport chain releasing energy, which allows hydrogen ions (H$^+$) to be pumped across the inner mitochondrial membrane. The return flow of these hydrogen ions (H$^+$) through the membrane protein **ATP synthase** results in the production of ATP. Finally, hydrogen ions (H$^+$) and electrons combine with **oxygen** to form water.

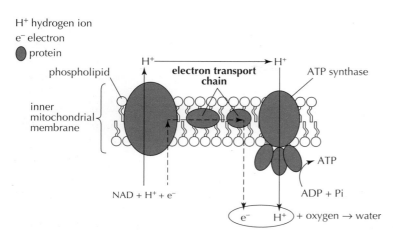

Figure 1.7.4 *The role of the electron transport chain in ATP synthesis*

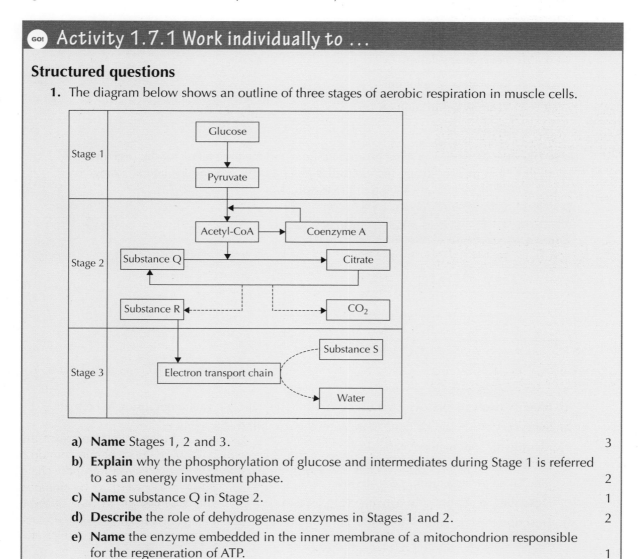

GO! Activity 1.7.1 Work individually to ...

Structured questions

1. The diagram below shows an outline of three stages of aerobic respiration in muscle cells.

   a) **Name** Stages 1, 2 and 3.   3
   b) **Explain** why the phosphorylation of glucose and intermediates during Stage 1 is referred
      to as an energy investment phase.   2
   c) **Name** substance Q in Stage 2.   1
   d) **Describe** the role of dehydrogenase enzymes in Stages 1 and 2.   2
   e) **Name** the enzyme embedded in the inner membrane of a mitochondrion responsible
      for the regeneration of ATP.   1

    f) **Name** substance S in Stage 3 and **describe** its role in aerobic respiration.   2

    g) **Describe** the role of ATP in cell metabolism.   1

2. The diagram represents a mitochondrion which has been magnified 10,000 times.

   Calculate the actual length of this mitochondrion in micrometres.   1

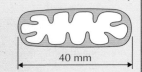

40 mm

3. During respiration in yeast, hydrogen ions (H^+) are released from glucose molecules. In experiments, these ions can decolourise resazurin, a blue indicator solution.

(1 mm = 1,000 micrometres)

Resazurin dye can therefore be used as an indicator of respiration and the time taken for resazurin to change colour will indicate the rate of respiration.

An investigation was carried out to study the effect of temperature on the rate of respiration in yeast. Four boiling tubes were set up as shown in the table and placed in a water bath at 20°C for 10 minutes.

Tube	Contents
1	5 cm³ yeast solution
2	5 cm³ glucose solution
3	5 cm³ resazurin solution
4	Empty

After 10 minutes, tubes 1–3 were mixed in tube 4 and the time taken for the resazurin to decolourise was measured and recorded. The procedure was then repeated at a range of temperatures. The results are shown in the table.

Temperature of water bath (°C)	Time for resazurin to decolourise (s)
20	120
25	90
30	60
35	45
40	30
45	90

  a) **Identify two** variables which must be controlled to allow a valid conclusion to be made.   2

  b) **Suggest** what should be done to increase the reliability of the procedure.   1

  c) **Identify** the independent variable.   1

  d) On a piece of graph paper, **draw a line graph** to show the effect of temperature on the time taken for resazurin to decolourise.   2

  e) **Describe** the relationship between temperature and the rate of respiration in yeast.   2

  f) **Describe** the contents of a suitable control tube to ensure that decolourisation was due to respiration.   1

(continued)

g) **Explain** why the four tubes were kept in the water bath for 10 minutes before their contents were mixed.   1

h) **Predict** how long it would have taken for resazurin to decolourise if the experiment had been repeated at 15°C.   1

> ### Make the link
>
> There is information about planning of variables, reliability and controls and presenting information as a line graph in Chapter 4 on page 263.

Extended response questions

1. **Give an account** of ATP synthesis in the electron transport chain on the inner membrane of the mitochondria.   4

2. **Give an account** of respiration under the following headings:

   a) Glycolysis   3

   b) The citric acid cycle   6

GO! Activity 1.7.2 Work in pairs to ...

1. **Practical activity: Investigating the activity of dehydrogenase in yeast using resazurin**

You will need: a marker pen, stopwatch, safety goggles, three test tubes, a syringe/Pasteur pipette, resazurin colour chart OR a colorimeter, 2·5% yeast solutions (live yeast and boiled yeast), 5% glucose, 0·01% resazurin dye, water bath at 35°C and distilled water.

During respiration in yeast, hydrogen ions (H^+) are released from glucose molecules. In experiments, these ions can decolourise resazurin, a blue indicator solution. Resazurin dye can therefore be used as an indicator of respiration and the time taken for resazurin to change colour will indicate the rate of respiration.

The resazurin goes through the following colour changes:

The numerical values for the various colours can be related to the respiratory (dehydrogenase) activity in yeast with 10 being equivalent to lowest activity.

⚠️ Your teacher will give specific safety instructions for carrying out this experiment and demonstrate any techniques needed.

- Wear eye protection while doing the experiment.
- Wash your hands with soap and water before and after doing the experiment.
- Ensure that all used apparatus and material is disposed of as instructed by your teacher.

Method

- Collect the materials indicated above.
- Label the three test tubes A, B and C and add 3 cm³ of resazurin dye to each tube.
- Add 3 cm³ of glucose solution to tubes A and C, and then add 3 cm³ of water to tube B.

- Add 3 cm^3 of live yeast suspension to tubes A and B, and then add 3 cm^3 of boiled yeast suspension to tube C.
- Gently shake each tube and place in a water bath at 35°C.
- Record the colour of each tube every 3 minutes for 30 minutes by comparing tubes with the colour chart or by using a colorimeter.

Results and report

You should each write a short report of what you did. Include an aim and a brief summary of the approach used to collect the experimental data. Remember to name any chemicals used and the equipment or method used to collect the data. Present your results in a table with clear headings.

Assignment Support

You could use this experimental technique to generate data for your assignment:

- You could investigate many aspects of the factors which affect respiration rates in yeast such as temperature, pH, different respiratory substrate, substrate concentration, concentration of yeast or effects of inhibitors.
- You could use a copy of the grid on page 288 to plan an assignment based on this experiment.

 2. Flashcard activity

You will each need: a set of blank flashcards (A7 cards) and a stopwatch.

- Find the glossary terms for this chapter – they are the **black** typeface and **red** typeface terms. Using your blank cards, you should each make a set of flashcards for these terms – write the term on one side and the definition on the other. You will find the definitions in the chapter.
- Shuffle your cards and lay them out in a column, some showing terms and some showing definitions – you decide. Your partner should match their cards with yours, laying their cards in a column beside yours to give the corresponding term or definition. Time how long they take to do this.
- Now swap roles – your partner should lay out their cards and you should try the matching exercise while your partner times you.
- You should each keep your set of flashcards as a revision tool for later.

GO! Activity 1.7.3 Work as a group to …

1. Dice and Slice: Respiration

You will need: Dice and Slice board (Appendix 2), the question and answer set and a dice.

- Take turns to play – have six turns each.
- Roll a dice for the top row number and again for the side number. One of the group will read you the question indicated by these numbers and you should try to answer it (they will tell you if you're right).
- If you get your question right, add your dice throw numbers to your score card and total them.
- After you have all had six turns you should find your own overall totals. Who has won?
- If you have time, play again and try to improve your scores.

2. Ring of Fire: Aerobic respiration

*You will need: a printout of the **Ring of Fire: Aerobic respiration** question and answer card set and a stopwatch.*

- Your teacher will deal the question and answer cards until all the cards have been issued.
- Your teacher will nominate a student to read the first question aloud and will start the clock.
- The student with the correct answer card should read the answer and then ask their own question.
- This is repeated until all the questions are completed. The timer is stopped.
- You should repeat the whole game to try to improve your time.

Learning checklist

After working on this chapter, I can:

Knowledge and understanding

1. State that glycolysis is the breakdown of glucose to pyruvate in the cytoplasm. ⬭ ⬭ ⬭

2. State that two ATP are required for the phosphorylation of glucose and intermediates during the energy investment phase of glycolysis. This leads to the generation of four ATP during the energy pay-off stage and results in a net gain of two ATP. ⬭ ⬭ ⬭

3. Explain that in aerobic conditions, pyruvate is broken down to an acetyl group that combines with coenzyme A to form acetyl coenzyme A. ⬭ ⬭ ⬭

4. Explain that in the citric acid cycle, the acetyl group from acetyl coenzyme A combines with oxaloacetate to form citrate. ⬭ ⬭ ⬭

5. Describe how during a series of enzyme-controlled steps, citrate is gradually converted back into oxaloacetate, which results in the generation of one ATP and release of carbon dioxide. ⬭ ⬭ ⬭

6. State that the citric acid cycle occurs in the matrix of the mitochondria.

7. State that dehydrogenase enzymes remove hydrogen ions (H^+) and electrons and pass them to the coenzyme NAD, forming NADH. This occurs in both glycolysis and the citric acid cycle.

8. State that the hydrogen ions (H^+) and electrons from NADH are passed to the electron transport chain on the inner membrane of the mitochondria.

9. Describe the electron transport chain as a series of carrier proteins attached to the inner mitochondrial membrane.

10. State that energy is released as electrons are passed along the electron transport chain.

11. State that energy from the flow of electrons allows hydrogen ions (H^+) to be pumped across the inner mitochondrial membrane.

12. State that the return flow of hydrogen ions (H^+) through the membrane protein ATP synthase results in the synthesis of ATP.

13. State that hydrogen ions (H^+) and electrons combine with oxygen to form water.

14. State that ATP is used to transfer energy to cellular processes which require energy such as cell division, DNA replication and protein synthesis.

Skills

1. *Process information to convert millimetres to micrometres.*

2. *Select information from a table.*

3. *Present data as a line graph.*

4. *Evaluate an experiment by identifying variables which should be controlled.*

5. *Evaluate an experiment by suggesting improvements to reliability and validity.*

6. *Plan an experiment by suggesting a suitable control.*

7. *Make a prediction from experimental data.*

1.8 Energy systems in muscle cells

You should already know:

- Glucose is broken down to two molecules of pyruvate, releasing enough energy to yield two molecules of ATP. Further breakdown depends upon the presence or absence of oxygen.
- If oxygen is present, aerobic respiration takes place and each pyruvate molecule is broken down to carbon dioxide and water, releasing enough energy to yield a large number of ATP molecules.
- In the absence of oxygen, the fermentation pathway takes place.
- In animal cells, the pyruvate molecules are converted to lactate.
- The breakdown of each glucose molecule via the fermentation pathway yields only the initial two molecules of ATP.
- Respiration begins in the cytoplasm and the process of fermentation is completed in the cytoplasm whereas aerobic respiration is completed in the mitochondria.

Learning intentions

- Describe lactate metabolism.
- Describe slow-twitch and fast-twitch muscle fibres and explain their role in different sports.

Lactate metabolism

During vigorous exercise, the muscle cells do not get sufficient oxygen to support the electron transport chain.

Under these conditions, pyruvate is converted to lactate. This conversion involves the transfer of hydrogen ions (H^+) from the NADH produced during glycolysis to pyruvate in order to produce lactate. This regenerates the NAD needed to maintain ATP production through glycolysis.

Lactate accumulates and **muscle fatigue** occurs. When exercise is complete, the **oxygen debt** is repaid. This allows respiration to provide the energy to convert lactate back to pyruvate and glucose in the liver. **Figure 1.8.1** provides a summary of lactate metabolism.

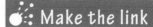 **Make the link**

There is more about the electron transport chain in Chapter 1.7 on page 75.

📖 **Muscle fatigue**

A decrease in the ability to produce force due to the accumulation of lactate.

📖 **Oxygen debt**

The amount of extra oxygen required by muscle tissue to convert lactate back into pyruvate and replenish depleted ATP.

Figure 1.8.2 *Skeletal muscles controlled by the somatic nervous system*

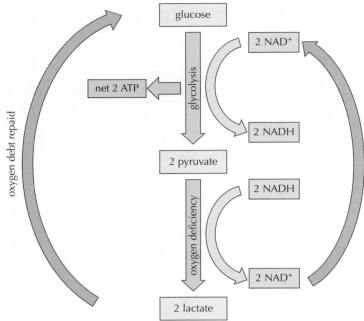

Figure 1.8.1 *Summary of lactate metabolism*

Make the link

There is more about the somatic nervous system in Chapter 3.1 on page 194.

Types of skeletal muscle fibres

Skeletal muscle is a form of muscle tissue which is under the voluntary control of the somatic nervous system.

Most skeletal muscles are attached to bones by tendons and their contraction results in muscle shortening and thus movement of the bone to which it is attached.

Skeletal muscle fibres can be subdivided into slow-twitch and fast-twitch fibres, with each having different properties, functions and distributions in the human body. Most human muscle tissue contains a mixture of both slow-twitch and fast-twitch muscle fibres. Athletes show distinct patterns of muscle fibres that reflect their sporting activities.

Slow-twitch muscle fibres

Slow-twitch muscle fibres rely on aerobic respiration to generate ATP and have many mitochondria.

Due to their large oxygen requirements, they have a large blood supply and a high concentration of the oxygen-storing protein **myoglobin**, which gives muscles their reddish colour. The major storage fuel of slow-twitch muscle fibres is fats. Slow-twitch muscle fibres contract relatively slowly but can sustain contractions for longer. They are useful for endurance activities such as long-distance running, cycling or cross-country skiing.

Hint

SS = Slow-twitch for Stamina

Myoglobin

An oxygen-binding protein found in the muscle tissue.

(a)

(b)

Figure 1.8.3 *Activities which rely on slow-twitch muscle fibres: (a) Long-distance cycling (b) Cross-country skiing*

Fast-twitch muscle fibres

Fast-twitch muscle fibres generate ATP through glycolysis alone and have fewer mitochondria, a lower blood supply and a lower concentration of the oxygen-storing protein myoglobin compared to slow-twitch muscle fibres. The major storage fuel of fast-twitch muscle fibres is **glycogen**. Fast-twitch muscle fibres contract relatively quickly, over short periods and are useful for activities such as sprinting or weightlifting.

The table below compares the characteristics of slow-twitch and fast-twitch muscle fibres.

(a)

(b)

Figure 1.8.4 *Activities which rely on fast-twitch muscle fibres: (a) Sprinting (b) Weightlifting*

Skeletal muscle fibres		
Characteristic	**Slow-twitch**	**Fast-twitch**
Generation of ATP	Aerobic respiration	Glycolysis only
Blood supply	Large blood supply	Low blood supply
Mitochondria	Many	Few
Concentration of myoglobin	High	Low
Major storage fuel	Fat	Glycogen
Rate of contraction	Contract slowly	Contract rapidly
Rate of fatigue	Fatigue occurs slowly	Fatigue occurs quickly
Type of activities	Endurance activities such as long-distance running, cycling or cross-country skiing.	High-intensity, short duration activities such as sprinting or weightlifting

🔍 **Hint**

FF = **F**ast-twitch for running **F**ast

📖 **Glycogen**

A major storage carbohydrate in fast twitch muscles.

💧 **Make the link**

There is more about glycolysis in Chapter 1.7 on page 75.

Activity 1.8.1 Work individually to ...

Structured questions

1. The diagram represents glycolysis and the metabolic pathway which synthesises lactate.

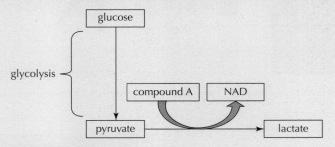

   a) **State** where glycolysis occurs within a cell. 1

   b) During lactate metabolism, NAD is regenerated.

      i. **Name** compound A. 1

      ii. **Explain** the importance of the regeneration of NAD for glycolysis. 1

      iii. **State** the reason why muscle cells produce lactate during vigorous exercise. 1

   c) **Give** the term used to describe the temporary shortage of oxygen arising from exercise. 1

2. **Describe two** structural differences between slow-twitch and fast-twitch muscle fibres. 2

3. **Choose** a sporting activity and decide whether slow-twitch or fast-twitch muscle fibres would be best suited for the activity.

   Give two reasons to justify your choice of muscle fibre. 3

4. The graph shows the percentage of slow- and fast-twitch muscle fibres present in athletes who trained for events of different distances.

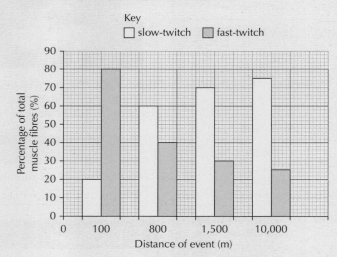

   a) **Describe** the relationship between the distance of the event trained for and the percentage of both types of muscle fibres. 2

   b) **Calculate**, as a simple whole number ratio, the percentage of slow-twitch muscle fibres to fast-twitch muscle fibres in athletes who have trained for the 10,000 m event. 1

(continued)

c) **Predict** the percentage of slow-twitch and fast-twitch muscle fibres that might be expected in an athlete training for a 400 m event.   1

 **Make the link**

There is information about ratios in Chapter 4 on page 269.

Extended response questions

1. **Write notes** on lactate metabolism in muscle cells.   4
2. **Give an account** of slow-twitch and fast-twitch muscle fibres.   6

GO! Activity 1.8.2 Work in pairs to …

1. **Card sort activity: Slow-twitch and fast-twitch muscle fibres**

   You will need: a copy of the card sort template (Appendix 1) and scissors.

   • Copy the statements in the table below into the spaces provided on the card sort template.
   • Cut the template into separate cards.
   • Shuffle the cards then sort the cards into two groups, one for slow-twitch muscle fibres and the other for fast-twitch muscle fibres.

Generate ATP through aerobic respiration	Suited for endurance activities such as long-distance running, cycling or cross-country skiing
Large blood supply	Low blood supply
Contract rapidly	Few mitochondria
High concentration of myoglobin	Lower concentration of myoglobin
Suited for high-intensity, short duration activities such as sprinting or weightlifting	Major storage fuel is fat
Many mitochondria	Slow to fatigue
Major storage fuel is glycogen	Quick to fatigue
Generate ATP by glycolysis alone	Contract slowly

2. **Card sort collage: Lactate metabolism**

   You will need: a card sort template (Appendix 1), whiteboard, Blu-Tack and a marker pen.

   • Copy the statements in the table below into the spaces provided on the card sort template.
   • Cut the template into separate cards.
   • Arrange the cards on the whiteboard in the correct pattern to show the events in lactate metabolism and use Blu-Tack to stick them down.
   • Use a marker pen to draw arrows linking the cards where required.

Glucose	NAD	Oxygen deficiency
NADH	Pyruvate	Lactate
Glycolysis	NADH	Oxygen debt repaid
NAD	Net gain of 2 ATP	

3. Flashcard activity

You will each need: a set of blank flashcards (A7 cards) and a stopwatch.

- Find the glossary terms for this chapter – they are the **black** typeface and red typeface terms. Using your blank cards, you should each make a set of flashcards for these terms – write the term on one side and the definition on the other. You will find the definitions in the chapter.

- Shuffle your cards and lay them out in a column, some showing terms and some showing definitions – you decide. Your partner should match their cards with yours, laying their cards in a column beside yours to give the corresponding term or definition. Time how long they take to do this.

- Now swap roles – your partner should lay out their cards and you should try the matching exercise while your partner times you.

- You should each keep your set of flashcards as a revision tool for later.

GO! Activity 1.8.3 Work as a group to …

1. **Design and make a poster or spider diagram detailing the characteristics and features of slow-twitch and fast-twitch muscle fibres.**

   You will need: A3 paper or card and coloured pencils, or a mini whiteboard and marker pens.

   Work as a group to design and make a poster or spider diagram. Include the activities for which each type of muscle fibre is useful.

Learning checklist

After working on this chapter, I can:

Knowledge and understanding

1. State that during vigorous exercise, the muscle cells do not receive sufficient oxygen to support the electron transport chain. Under these conditions, pyruvate is converted to lactate.

2. State that the conversion of pyruvate to lactate involves the transfer of hydrogen from the NADH produced during glycolysis to pyruvate in order to produce lactate.

3. Explain that the regeneration of NAD is needed to maintain ATP production through glycolysis.

4. State that the accumulation of lactate causes muscle fatigue to occur.

5. State that the oxygen debt is repaid when exercise is complete, which allows respiration to provide the energy required to convert lactate back to pyruvate and glucose in the liver.

6. State that most human muscle tissue contains a mixture of both slow-twitch and fast-twitch muscle fibres.

7. State that slow-twitch muscle fibres contract relatively slowly but can sustain contractions for longer. They are useful for endurance activities such as long-distance running, cycling or cross-country skiing.

8. State that slow-twitch muscle fibres rely on aerobic respiration to generate ATP and have many mitochondria, a large blood supply and a high concentration of the oxygen-storing protein myoglobin.

9. State that the major storage fuel of slow-twitch muscle fibres is fats.

10. State that fast-twitch muscle fibres contract relatively quickly, over short periods.

11. State that fast-twitch muscle fibres are useful for activities such as sprinting or weightlifting.

12. State that fast-twitch muscle fibres can generate ATP through glycolysis only.

Skills

1. *Process information to calculate a simple whole number ratio.*

2. *Select information from a bar chart.*

3. *Present data as a line graph.*

4. *Make predictions based upon trends in data.*

Chapter 1 practice area test
Human cells

Write your answers on separate sheets of paper. Mark your work using the answers online at www.collins.co.uk/pages/Scottish-curriculum-free-resources.

Paper 1: Multiple choice

Total: 10 marks

1. Which pathway describes the production of haploid gametes from diploid germline cells?

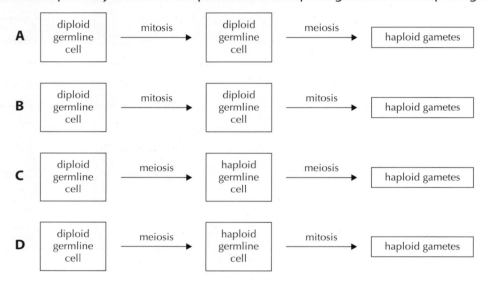

2. The list shows steps in the polymerase chain reaction (PCR).

   1. Binding of a primer to DNA

   2. Action of DNA polymerase

   3. Sample of DNA heated

   4. Separation of DNA strands

   In which sequence do these steps occur?

   A 1, 2, 4, 3

   B 1, 2, 3, 4

   C 3, 4, 1, 2

   D 3, 4, 2, 1

3. DNA profiling may be carried out on samples amplified by PCR and used to solve crimes. During this procedure, DNA from the amplified sample is cut into fragments by enzymes which cut the DNA at specific points. The diagram shows a DNA sample being cut by two different enzymes.

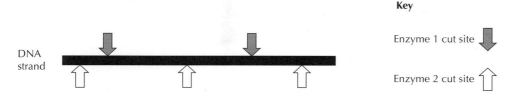

Which line in the table shows the number of fragments which would be produced by the use of different combinations of the two enzymes?

	Number of fragments produced		
	Using Enzyme 1 only	*Using Enzyme 2 only*	*Using both Enzymes 1 and 2*
A	2	3	5
B	2	3	6
C	3	4	7
D	3	4	6

4. The diagram identifies three stages, X, Y and Z, which occur during gene expression.

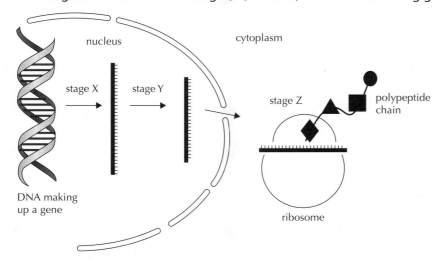

Which row in the table identifies these three stages?

	Stage X	*Stage Y*	*Stage Z*
A	RNA splicing	Transcription	Translation
B	Translation	Transcription	RNA splicing
C	Transcription	RNA splicing	Translation
D	Transcription	Translation	RNA splicing

5. The list shows three applications of technology in human genomics.

    1. Comparing genomes to help solve crimes

    2. Diagnosing genetic disease through genome screening

    3. Selecting drug treatments based on genome sequencing

Which of these applications are examples of pharmacogenetics?

    A   1 only

    B   3 only

    C   2 and 3 only

    D   1, 2 and 3

6. The diagram shows the molecular shapes of an enzyme and its product molecule.

Enzyme molecule   *Product molecule*

Which line in the table shows the possible molecular shapes of different inhibitors of this enzyme?

	Competitive inhibitor	Non-competitive inhibitor
A	●	⬠
B	▲	■
C	■	●
D	⏣	■

7. The diagram shows a metabolic pathway in a human cell.

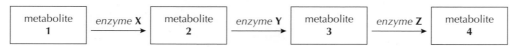

In end-product inhibition

    A   enzyme Z binds to enzyme X

    B   enzyme Z binds to metabolite 1

    C   metabolite 4 binds to enzyme X

    D   metabolite 4 binds to metabolite 1

8. During glycolysis, dehydrogenase enzymes catalyse the

   A   removal of hydrogen ions (H^+) from $NADH_2$

   B   removal of hydrogen ions (H^+) from citrate

   C   transfer of hydrogen ions (H^+) to glucose

   D   transfer of hydrogen ions (H^+) to NAD

9. The list shows some of the substances produced during the respiration of a molecule of glucose in the presence of oxygen.

   1.   acetyl group

   2.   pyruvate

   3.   citrate

   4.   ATP

   In which order would these substances appear?

   A   4, 2, 1, 3

   B   4, 2, 3, 1

   C   2, 1, 4, 3

   D   2, 3, 1, 4

10. During fermentation in human cells, pyruvate is converted to lactate.

    This reaction is useful for metabolism because it

    A   avoids the production of carbon dioxide

    B   produces large amounts of ATP

    C   regenerates NAD from NADH

    D   recycles ADP for use in glycolysis

Paper 2: Structured and extended response

Total: 40 marks

1. The diagram shows the role of embryonic stem cells in the development of a human embryo.

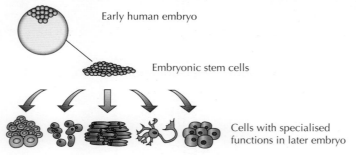

Early human embryo

Embryonic stem cells

Cells with specialised functions in later embryo

  a) i.   **Give** the term used to describe the process by which a cell develops specialised functions.   1

       ii. **Explain** how an embryonic stem cell can develop into a specialised cell.   1

  b) **Give** the meaning of the term 'multipotent' as applied to tissue stem cells.   1

  c) **Describe** how cancer cells form a tumour and explain how a secondary tumour develops.   2

2. The diagram below shows part of a DNA molecule during replication.

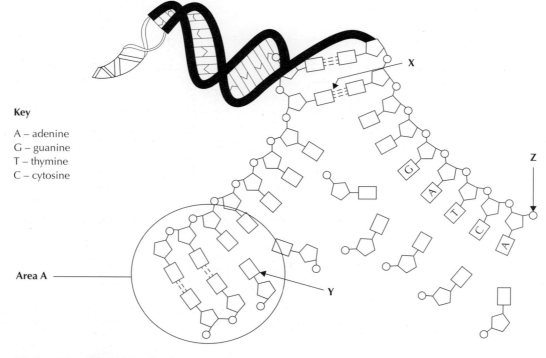

Key

A – adenine
G – guanine
T – thymine
C – cytosine

X

Z

Area A

Y

  a) i.   **Name** the bonds shown at X.   1

       ii. **Identify** the nucleotide base at Y.   1

       iii. **Name** the component of the DNA backbone at Z.   1

b) DNA is replicated continuously on the leading strand of the parent DNA as shown in Area A.

Describe replication on the other strand of the parent DNA. **1**

c) **Explain** why DNA must replicate. **1**

3. a) The diagram shows part of a DNA template strand and a part of a primary RNA transcript synthesised from it.

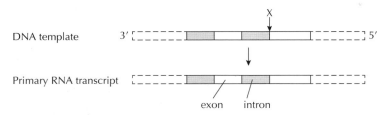

i. **Give** the term used to describe the process shown in the diagram. **1**

ii. Using information in the diagram, **describe** a possible effect on the primary RNA transcript of a single nucleotide mutation at point X on the DNA template. **1**

b) DNA is encoded in triplet sequences.

Explain what is meant by this. **1**

c) A translocation chromosome mutation produces substantial changes to an affected individual's genetic material.

i. **Describe** what is meant by 'translocation'. **1**

ii. Apart from translocation, **name one** other type of chromosome mutation which can affect the structure of human chromosomes. **1**

4. Hydrogen peroxide is a toxic chemical which is produced in human metabolism.

Catalase is an enzyme that breaks down hydrogen peroxide as shown below.

$$\text{hydrogen peroxide} \xrightarrow{\text{catalase}} \text{water + oxygen}$$

Experiments were carried out to investigate how a competitive inhibitor affected the rate of this reaction.

Ten filter paper discs were soaked in a 5% catalase solution and then each disc was added to a beaker of hydrogen peroxide solution to which a different concentration of the inhibitor had previously been added, as shown in the diagram.

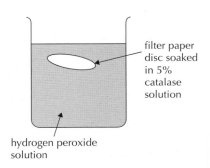

filter paper disc soaked in 5% catalase solution

hydrogen peroxide solution

The discs sank to the bottom of the beaker before rising back up to the surface. The time taken for each disc to rise to the surface was used to measure the reaction rates. The results of the investigation are shown in the table.

Inhibitor concentration (%)	Average time for 10 discs to rise (s)
0·1	1·8
0·5	3·2
1·0	5·6
1·5	7·2
2·0	9·4
2·5	9·4

a) **Explain** why the filter paper discs rose to the surface of the hydrogen peroxide solution. **1**

b) i. **Give** the dependent variable in this experiment. **1**

   ii. **Give one** variable that should be controlled during this investigation. **1**

c) **Describe one** feature of this investigation which makes the results more reliable. **1**

d) **Plot a line graph** to show the results of the investigation. **2**

e) **Give two** conclusions that can be drawn from these results. **2**

5. The diagram shows part of the electron transport chain in human cells.

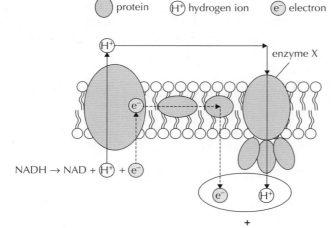

a) **Give** the exact location of the electron transport chain in human cells. **1**

b) In this system, hydrogen ions (H^+) are pumped across the inner mitochondrial membrane as shown in the diagram.

   Describe the source of energy required for this process. **1**

c) The return flow of hydrogen ions (H^+) drives enzyme X, which results in the production of ATP.

Name enzyme X. **1**

d) **Name** the final hydrogen acceptor molecules and the final metabolic product shown in the diagram. **2**

6. The chart below shows how the average percentages of slow-twitch and fast-twitch fibres found in the muscles of athletes varies with the distance of the event for which they train.

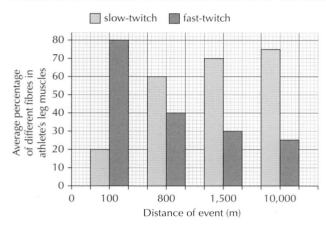

a) **i.** **Compare** the trends shown in the chart. **1**

ii. **Express** the percentage of slow-twitch to fast-twitch fibres found in the muscles of the 10,000 m runner as a simple whole number ratio. **1**

b) **Describe one** functional and **one** structural difference between slow- and fast-twitch fibres. **2**

7. **Describe** the mode of action of enzymes in the control of metabolic pathways under the following headings:

a) Activation energy and induced fit **4**

b) Effects of substrate concentration **4**

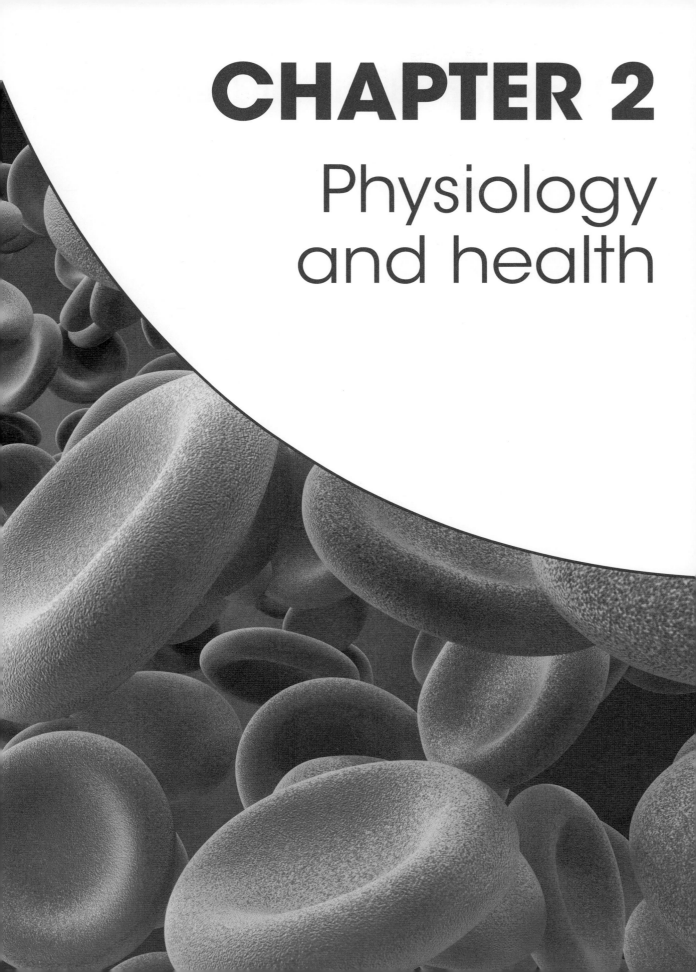

CHAPTER 2

Physiology and health

2.1 Gamete production and fertilisation

You should already know:

- Body cells are diploid, except gametes, which are haploid.
- The types of gametes, the organs that produce them, and where these are located in animals. The basic structure of sperm and egg cells.
- Fertilisation is the fusion of the nuclei of the two haploid gametes to produce a diploid zygote, which divides to form an embryo.

Learning intentions

- Describe gamete production in the testes and the role of the prostate gland and seminal vesicles.
- Describe gamete production in the ovaries.
- Describe the process of fertilisation.

Gamete production

🔅 Make the link

There is more about germline cells and meiosis in Chapter 1.1 on page 12.

Gametes (sperm and ova) are produced from the germline cells found in the male and female reproductive organs. Germline cells are found in the ovaries of females and the testes of males. They can divide by mitosis to form more germline cells and can also divide by another type of cell division called meiosis, which results in the formation of the gametes. Gametes are haploid, which means that they each contain a single set of chromosomes.

📖 Seminiferous tubules

Narrow coiled tubes in the testes that produce sperm.

Gamete production in the testes

Figure 2.1.1 shows features of the male reproductive system.

📖 Interstitial cells

Cells between the seminiferous tubules in the testes that produce testosterone.

📖 Testosterone

Primary male sex hormone produced by the interstitial cells.

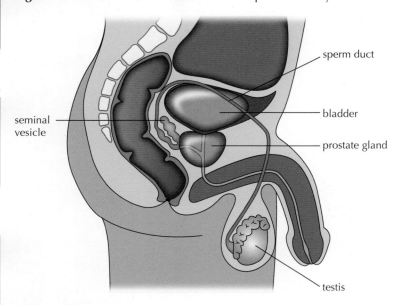

Figure 2.1.1 *The male reproductive system*

The testes are the male reproductive organs. Each testis contains several coiled **seminiferous tubules** within which germline cells divide, first by mitosis and then by meiosis, to produce the haploid sperm cells. The **interstitial cells**, found between the seminiferous tubules of the testes, produce the hormone **testosterone**. Testosterone stimulates sperm production in the seminiferous tubules and also activates the **prostate gland** and **seminal vesicles** to secrete fluids which, when added to the sperm, are collectively called seminal fluid or **semen**. The fluid produced by the prostate gland and seminal vesicles maintains the **mobility** and **viability** of the sperm by providing a liquid at the optimum viscosity for the sperm to swim in and sugar which provides the sperm cells with an energy source. **Figure 2.1.2** shows a section through a testis and seminiferous tubule.

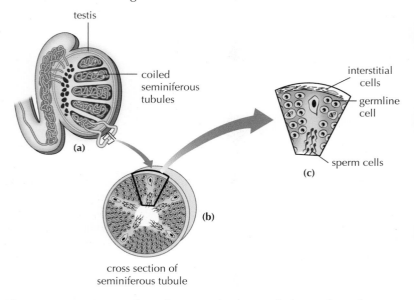

Figure 2.1.2 *(a) Section through a testis (b) a magnified view through a section of a seminiferous tubule and (c) a further magnified view of cells in the tubule and the position of interstitial cells*

Gamete production in the ovaries

The structure of the female reproductive system is shown in **Figure 2.1.3**.

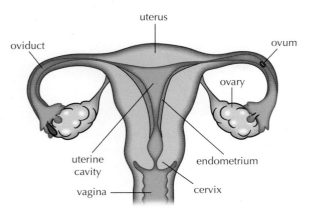

Figure 2.1.3 *The structure of the female reproductive system*

Prostate gland
Produces fluid that helps maintain the mobility and viability of the sperm in the semen.

Seminal vesicles
Produces fluid that helps maintain the mobility and viability of the sperm in the semen.

Semen
The male reproductive fluid containing the sperm.

Mobility
The ability of sperm to move properly through the female reproductive tract to reach the egg.

Viability
Refers to the percentage of live sperm in the semen sample.

Make the link
There is more about testosterone, the prostate gland and the seminal vesicles in Chapter 2.2 on page 107.

Hint
Remember – in**T**ers**T**i**T**ial cells produce **T**es**T**os**T**erone.

Follicle
Fluid - filled sac of cells that protects the developing ovum and produces oestrogen.

📖 Oestrogen

Hormone secreted by the follicle that repairs and thickens the endometrium.

📖 Ovulation

Release of ovum from a mature follicle.

📖 Corpus luteum

Mass of cells formed from the follicle after ovulation that produces progesterone.

📖 Progesterone

Hormone secreted by the corpus luteum that continues to thicken and vascularise the endometrium.

⚛ Make the link

There is more about oestrogen and progesterone in Chapter 2.2 on page 107 and Chapter 2.3 on page 118.

⚛ Make the link

There is more about the corpus luteum in Chapter 2.2 on page 110.

📖 Endometrium

The innermost lining of the uterus and the layer in which implantation takes place.

⚛ Make the link

There is more about the endometrium in Chapter 2.2 on page 109.

The ovaries are the female reproductive organs. At birth, the ovaries contain approximately one million immature follicles formed by the germline cells. Once a female begins ovulating at puberty, an ovum matures inside a **follicle** approximately every 28 days. Each ovum is surrounded by a follicle, which protects the developing ovum and secretes the hormone **oestrogen**. After **ovulation**, the ovum passes into the oviduct where it may be fertilised by a sperm cell to form a diploid zygote. The follicle then develops into a **corpus luteum** ('yellow body'), which now secretes the hormone **progesterone**. **Figure 2.1.4** shows some features of the ovary, maturation of follicle, ovulation and the development and degeneration of the corpus luteum.

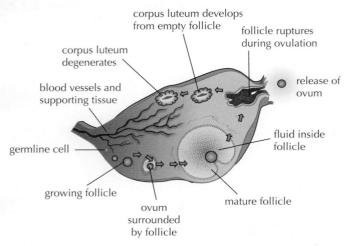

Figure 2.1.4 *Section through an ovary showing the development of an ovum from a germline cell*

Fertilisation

Mature ova are released into the oviduct where they may be fertilised by sperm to form a zygote. The zygote then starts to divide by mitosis to form a small ball of cells called an embryo that implants into the **endometrium** of the uterus as shown in **Figure 2.1.5**.

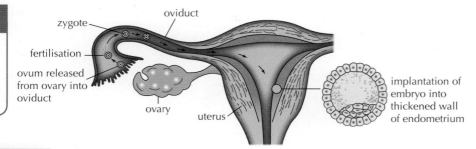

Figure 2.1.5 *Release of ovum into oviduct followed by fertilisation and implantation of the embryo into the thickened endometrium*

GO! Activity 2.1.1 Work in pairs to ...

Structured questions

1. A section through part of a testis is shown in the diagram below.

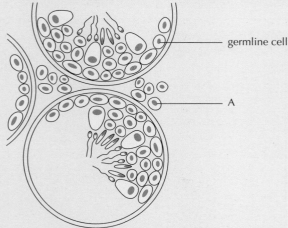

germline cell

A

   a) **Give** the name of the tubules shown in the diagram.   1
   b) **Describe** the role of the germline cells in the testes.   2
   c) **State** the name of the cells labelled A that secrete the hormone testosterone.   1
   d) **Describe two** functions of testosterone.   2
2. **Describe** the role of the prostate gland and seminal vesicles.   2

⚗️ Make the link

There is more on processing information in Chapter 4 on page 269.

3. The diagrams show sections through two structures found at different times in the ovary.

  P

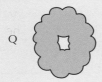 Q

   a) i.   **Give** the name of structure P that surrounds and protects the developing ovum.   1
      ii. **Name** structure Q that develops after ovulation.   1
      iii. **State** which event in the menstrual cycle occurs before P develops into Q.   1
   b) **State one** other function of structures P and Q.   1
4. **Give** the location of fertilisation in the female reproductive system.   1

(Continued)

5. During a study which aimed to explain the decreasing birth rates in a small European country, investigators looked at historical data on the average sperm counts from groups of men in the country between 1940 and 2000. The results are shown in the graph.

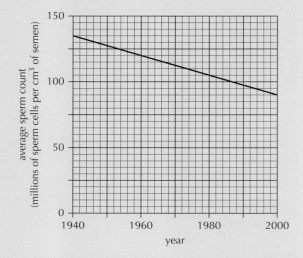

a) **Suggest three** features of the selection of groups of men which the investigators should have checked to evaluate the validity **and** reliability of their results.   3

b) i. **Calculate** the average reduction in sperm count per year.   1

   ii. **Calculate** the average number of sperm cells which would be expected in a $2\,cm^3$ sample of semen taken in 1990.   1

c) **Identify** the years between which the average sperm count decreased from 125 to 105 million cells per cm^3 of semen.   1

d) **Give a conclusion** which the investigators may reach.   1

Extended response questions

1. **Give an account** of gamete production in the testes and the role of the prostate gland and seminal vesicles.   6

2. **Describe** the development of ova in the ovary.   4

Activity 2.1.2 Work in pairs to ...

1. **Practical activity: Examining a section through an ovary**

    You will need: a microscope and a prepared slide of ovarian tissue (may not be from a human source but the basic structure is similar). You will each need a sharp pencil and a piece of A4 blank paper.

    ⚠️ Your teacher will give specific safety instructions for carrying out this experiment and demonstrate any techniques needed. Ensure that all used apparatus and material is disposed of as instructed by your teacher.

    Method

    *   Set up your microscope as directed by your teacher.
    *   Place the prepared slide of ovarian tissue onto the stage and focus at high magnification on the preparation. If you can't do this, use the photograph below.
    *   Look at the individual cells.
    *   Try to identify different stages of follicular development (immature follicle and mature follicle) and a corpus luteum.

    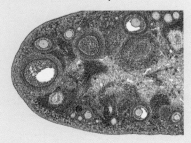

    a) Using a piece of A4 blank paper and a sharp pencil, draw and label a diagram of your slide or the photograph, showing the developing follicles and possibly the corpus luteum. Add your diagram into your notes.

    b) Answer the following questions:

        i.   **Name** the tissue from which the preparation comes.   1
        ii. **Name** the gametes which are being made here.   1
        iii. **Name** the structure that develops from the follicle after ovulation.   1

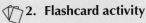 2. **Flashcard activity**

    You will each need: a set of blank flashcards (A7 cards) and a stopwatch.

    *   Find the glossary terms for this chapter – they are the **black** typeface and **red** typeface terms. Using your blank cards, you should each make a set of flashcards for these terms – write the term on one side and the definition on the other. You will find the definitions in the chapter.
    *   Shuffle your cards and lay them out in a column, some showing terms and some showing definitions – you decide. Your partner should match their cards with yours, laying their cards in a column beside yours to give the corresponding term or definition. Time how long they take to do this.
    *   Now swap roles – your partner should lay out their cards and you should try the matching exercise while your partner times you.
    *   You should each keep your set of flashcards as a revision tool for later.

GO! Activity 2.1.3 Work as a group to ...

1. **Placemat activity: Gamete production**

   You will need: a placemat template (Appendix 3) and four fine marker pens.

   - Set the placemat in the middle of the table and each write your name in a section.
   - Each participant should then write words that they think are related to *gamete production in the testes and ovaries* into their section of the placemat. Spend 2 minutes doing this.
   - You should take it in turns to read out a word from your section. If everyone agrees it is related to the topic it can be copied into the centre section of the placemat. Continue until all words have been discussed.
   - Working as a group, use all the words in the centre section to summarise your knowledge of gamete production in the testes and ovaries.

Learning checklist

After working on this chapter, I can:

Knowledge and understanding

1. State that gametes are produced from germline cells by meiosis.

2. State that sperm are produced in the seminiferous tubules in the testes.

3. State that the interstitial cells of the testes produce the hormone testosterone.

4. State that the prostate gland and seminal vesicles secrete fluids that maintain the mobility and viability of the sperm.

5. State that the ovaries contain immature ova (egg cells) in various stages of development.

6. State that each ovum is surrounded by a follicle that protects the developing ovum and secretes hormones.

7. State that mature ova are released into the oviduct where they may be fertilised by sperm to form a zygote.

Skills

1. *Evaluate reliability and validity.*

2. *Process information to calculate an average decrease.*

3. *Process information through general calculation.*

4. *Select information from a line graph.*

5. *Draw a conclusion from a line graph.*

2.2 Hormonal control of reproduction

You should already know:

- Endocrine glands release hormones into the bloodstream.
- Hormones are chemical messengers.
- Target tissues have cells with complementary receptor proteins for specific hormones, so only that tissue will be affected by these hormones.
- Testes and ovaries are endocrine glands.

Learning intentions

- Describe the influence of hormones on puberty.
- Describe the hormonal control of sperm production.
- Describe the hormonal control of the menstrual cycle.

Hormones and reproduction

The control of reproduction depends on communication between tissues and organs that are widely separated. The chemical signals carrying out this communication are hormones, which are released into the blood and detected by their target cells through receptors on, or inside, their cell membranes. The reproductive hormones control the production of gametes at puberty and their continued production throughout the fertile life of an individual.

Hormonal influence on puberty

Puberty is the time in life when a boy or girl becomes sexually mature and capable of reproduction. It is a process that usually happens between ages 10 and 14 for girls and ages 12 and 16 for boys.

At puberty, the hypothalamus in the brain secretes a **releaser hormone** that targets the pituitary gland at the base of the brain. The pituitary gland is stimulated to release **follicle stimulating hormone (FSH)**, **luteinising hormone (LH)** in girls or **interstitial cell stimulating hormone (ICSH)** in boys. The release of these hormones triggers the onset of puberty, bringing about the onset of the **menstrual cycle** in females and the production of sperm in males.

Hormonal control of sperm production

In males, follicle stimulating hormone (FSH) promotes sperm production in the seminiferous tubules and interstitial cell stimulating hormone (ICSH) stimulates the interstitial cells in the testes to secrete testosterone. Along with FSH, testosterone also stimulates sperm production and activates the prostate gland and seminal vesicles to secrete fluids that maintain the mobility and viability of the sperm.

📖 Releaser hormone
Hormone released by the hypothalamus which affects the pituitary gland to trigger puberty.

📖 Follicle stimulating hormone (FSH)
Pituitary hormone that controls the development of follicles in the ovaries and sperm production in males.

📖 Luteinising hormone (LH)
Pituitary hormone that triggers ovulation and the development of the corpus luteum.

📖 Interstitial cell stimulating hormone (ICSH)
Pituitary hormone that stimulates testosterone production by the interstitial cells in the testes.

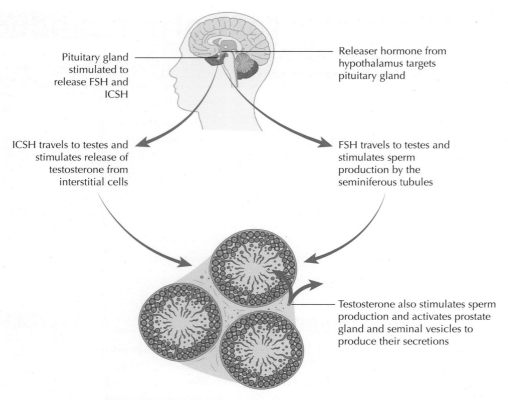

Pituitary gland stimulated to release FSH and ICSH

Releaser hormone from hypothalamus targets pituitary gland

ICSH travels to testes and stimulates release of testosterone from interstitial cells

FSH travels to testes and stimulates sperm production by the seminiferous tubules

Testosterone also stimulates sperm production and activates prostate gland and seminal vesicles to produce their secretions

Figure 2.2.1 *Hormonal control of sperm production in the testes*

Negative feedback control of testosterone by FSH and ICSH

Negative feedback is a biological control system in which the end product of a process reduces the stimulus of that same process, as when a high level of a particular hormone in the blood may inhibit further secretion of that hormone.

In males, high levels of testosterone inhibit the production of both FSH and ICSH by the pituitary gland. This leads to a decrease in the testosterone produced by the interstitial cells. When the testosterone levels in the blood drop to a certain level, it no longer inhibits the production of FSH and ICSH, which then increase once again. Negative feedback control ensures that both testosterone levels and sperm production remain relatively constant. This control is shown in **Figure 2.2.2**.

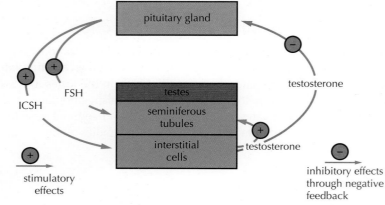

Figure 2.2.2 *Negative feedback control by testosterone on FSH and ICSH production*

stimulatory effects

inhibitory effects through negative feedback

Hormonal control of the menstrual cycle

In females, the menstrual cycle takes approximately 28 days with the first day of **menstruation** or a period regarded as day 1 of the cycle. It can be divided into two phases, as shown in **Figure 2.2.3**.

> **📖 Menstrual cycle**
>
> An approximately 28-day cycle in the middle of which ovulation occurs.

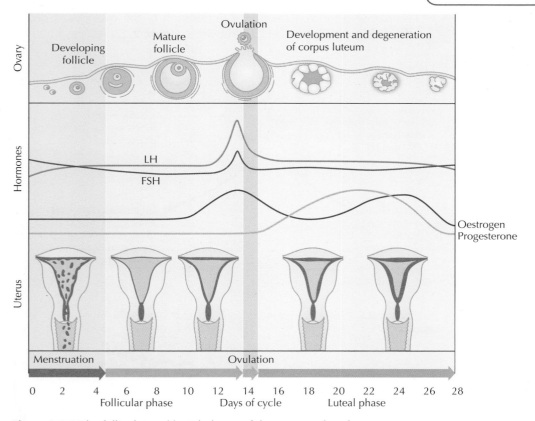

Figure 2.2.3 *The follicular and luteal phases of the menstrual cycle*

Follicular phase

In the **follicular phase** of the menstrual cycle, follicle stimulating hormone (FSH) released by the pituitary gland stimulates the development of a follicle in the ovary and the production of the hormone oestrogen by the follicle.

Oestrogen released by the developing follicle stimulates proliferation (thickening) of the endometrium, preparing it for implantation, and affects the consistency of cervical mucus, making it more watery and more easily penetrated by sperm. Peak levels of oestrogen stimulate a surge in the secretion of luteinising hormone (LH). This surge in LH triggers ovulation. Ovulation is the release of an egg (ovum) from a follicle in the ovary. It usually occurs around the midpoint (day 14) of the menstrual cycle.

> **📖 Menstruation**
>
> Removal of the endometrium and an unfertilised egg cell at the end of a menstrual cycle.

> **🔍 Hint**
>
> Remember **FF** – **F**SH is the **F**irst of the pituitary hormones released in the cycle.

> **📖 Follicular phase**
>
> The first stage in the menstrual cycle during which a follicle develops.

📖 Luteal phase

The second stage of the menstrual cycle in which a corpus luteum is present.

🔍 Hint

Remember **LL** – **L**H is the **L**ast of the pituitary hormones released in the cycle.

Luteal phase

In the **luteal phase** of the menstrual cycle, the ruptured follicle which released the ovum during ovulation develops into a corpus luteum, which secretes the hormone progesterone. Progesterone promotes further development and vascularisation of the endometrium, preparing it for implantation if fertilisation of the ovum does occur. The effects of FSH and LH are shown in **Figure 2.2.4**.

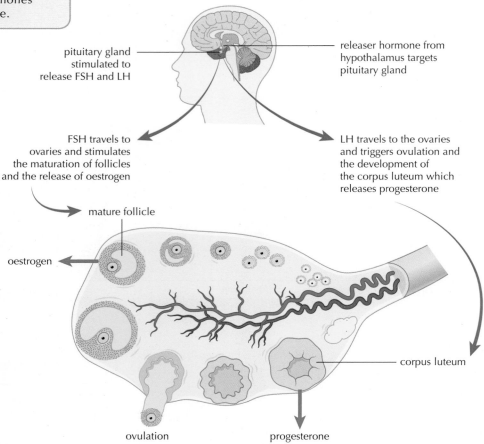

pituitary gland stimulated to release FSH and LH

releaser hormone from hypothalamus targets pituitary gland

FSH travels to ovaries and stimulates the maturation of follicles and the release of oestrogen

LH travels to the ovaries and triggers ovulation and the development of the corpus luteum which releases progesterone

mature follicle

oestrogen

corpus luteum

ovulation

progesterone

Figure 2.2.4 *The hormones of the menstrual cycle*

🔍 Hint

The ovarian hormones are released alphabetically: **O**estrogen then **P**rogesterone, **O** before **P**.

Negative feedback control of the menstrual cycle

FSH and LH production form part of a negative feedback cycle with oestrogen and progesterone.

The high levels of oestrogen and progesterone inhibit the pituitary gland from secreting FSH and LH.

The low level of FSH suppresses the development of any further follicles. The low level of LH causes the corpus luteum to degenerate and progesterone secretion to fall to a minimum. The falling level of progesterone at the end of the cycle triggers the start of menstruation with the loss of the inner layer of the endometrium, along with some blood.

The low level of oestrogen at the end of the cycle causes the pituitary gland to increase the secretion of FSH, which restarts the cycle. However, if fertilisation does occur, a hormone from the implanted embryo mimics the effect of LH and causes the corpus luteum to continue producing progesterone for another eight weeks until this function is taken over by the placenta. **Figure 2.2.5** summarises the negative feedback control of the menstrual cycle.

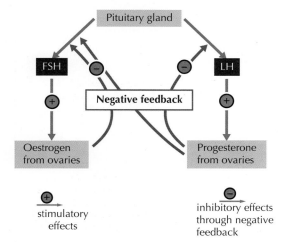

stimulatory effects

inhibitory effects through negative feedback

Figure 2.2.5 *Negative feedback control by oestrogen and progesterone on FSH and LH production*

> ● **Make the link**
>
> There is more about negative feedback and the role of oestrogen and progesterone in contraceptive pills in Chapter 2.3 on page 123.

GO! Activity 2.2.1 Work individually to ...

Structured questions

1. The flowchart summarises the hormonal control of semen production in humans.

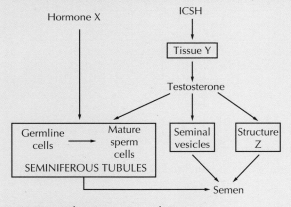

   a) **Name** hormone X and tissue Y. 2

   b) Semen contains substances secreted by structure Z.

      i. **Identify** structure Z. 1

      ii. **Describe two** functions of testosterone. 2

      iii. **Describe** how negative feedback control raises the concentration of testosterone in the blood if it has fallen to a low level. 2

   c) **Describe** the role of the secretions from the seminal vesicles and structure Z. 2

   d) **Name** the type of cell division responsible for the production of haploid sperm cells from the diploid germline cells in the seminiferous tubules. 1

2. Changes in the ovary during the menstrual cycle are described below.

    1 Corpus luteum forms

    2 Ovulation occurs

    3 Progesterone is produced

    4 Corpus luteum degenerates

    5 Follicle develops

    Give the sequence in which these changes would occur following menstruation.   1

3. The graph shows how the concentration of oestrogen in the blood and the thickness of the endometrium of the uterus vary during a menstrual cycle.

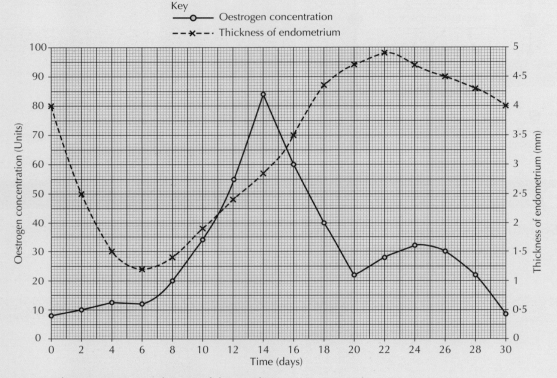

Key
— o — Oestrogen concentration
- - x - - Thickness of endometrium

a) Ovulation occurs on day 15 of this cycle.

    i. **Describe** the role of oestrogen in triggering ovulation.   2

    ii. **State** the concentration of oestrogen when the thickness of the endometrium was 4·8 mm.   1

b) **Express**, as a simple whole number ratio, the thickness of the endometrium on day 6 compared to day 22.   1

c) **Calculate** the percentage increase in the oestrogen concentration between the start of the cycle and day 14.   1

d) Oestrogen stimulates proliferation of the endometrium.

    Describe evidence from the graph that indicates another factor also stimulates thickening of the endometrium.   1

e) **Suggest one** way in which the graph for the next menstrual cycle would differ if the woman became pregnant during that cycle.   1

> ** Make the link**
>
> There is information about selecting and processing information in Chapter 4 on page 267.

Extended response questions

1. **Give an account** of hormonal influence on puberty.   4
2. **Describe** hormonal control of the menstrual cycle under the following headings:
   a) Events leading to ovulation   6
   b) Events following ovulation   4

GO! Activity 2.2.2 Work in pairs to . . .

 1. Flashcard activity

You will each need: a set of blank flashcards (A7 cards) and a stopwatch.

- Find the glossary terms for this chapter – they are the **black** typeface and **red** typeface terms. Using your blank cards, you should each make a set of flashcards for these terms – write the term on one side and the definition on the other. You will find the definitions in the chapter.

- Shuffle your cards and lay them out in a column, some showing terms and some showing definitions – you decide. Your partner should match their cards with yours, laying their cards in a column beside yours to give the corresponding term or definition. Time how long they take to do this.

- Now swap roles – your partner should lay out their cards and you should try the matching exercise while your partner times you.

- You should each keep your set of flashcards as a revision tool for later.

GO! Activity 2.2.3 Work as a group to . . .

 1. Dice and Slice: Hormonal control of reproduction

You will need: Dice and Slice board (Appendix 2), the question and answer set and a dice.

- Take turns to play – have six turns each.

- Roll a dice for the top row number and again for the side number. Your partner will read you the question indicated by these numbers and you should try to answer it (your partner will tell you if you're right).

- If you get your question right, add your dice throw numbers to your score card and total them.

- After you have both had six turns you should find your own overall totals. Who has won?

- If you have time, play again and try to improve your scores.

2. Ring of Fire: Hormonal control of reproduction

*You will need: a printout of the **Ring of Fire: Hormonal control of reproduction** question and answer card set and stopwatch.*

- Your teacher will deal the question and answer cards until all the cards have been issued.

- Your teacher will nominate a student to read the first question aloud and will start the clock.

(Continued)

- The student with the correct answer card should read the answer and then ask their own question.
- This is repeated until all the questions are completed. The timer is stopped.
- You should repeat the whole game to try to improve your time.

Learning checklist

After working on this chapter, I can:

Knowledge and understanding

1. State that hormones are chemical messengers produced by the endocrine glands.

2. State that hormones are released directly into the bloodstream and travel to their target tissue or organ where they have their effect.

3. State that hormones control the onset of puberty, sperm production and the menstrual cycle.

4. State that at puberty, the hypothalamus in the brain secretes a releaser hormone that targets the pituitary gland.

5. State that the pituitary gland is stimulated to release one hormone called follicle stimulating hormone (FSH), and a second hormone called luteinising hormone (LH) in girls and interstitial cell stimulating hormone (ICSH) in boys and that these hormones trigger the onset of puberty.

6. State that in males, FSH promotes sperm production in the seminiferous tubules of the testes and ICSH stimulates the interstitial cells in the testes to produce the male sex hormone testosterone. Testosterone also stimulates sperm production and activates the prostate gland and seminal vesicles to produce their fluid secretions.

7. State that the overproduction of testosterone is prevented by a negative feedback mechanism.

8. State that high testosterone levels inhibit the secretion of FSH and ICSH from the pituitary gland, resulting in a decrease in the production of testosterone by the interstitial cells.

9. State that the menstrual cycle takes approximately 28 days with the first day of menstruation regarded as day 1 of the cycle.

10. State that the pituitary hormones FSH and LH and the ovarian hormones oestrogen and progesterone are associated with the menstrual cycle.

11. State that in the follicular phase (first half of the cycle), FSH stimulates the development and maturation of a follicle surrounding the ovum and the production of the sex hormone oestrogen by the follicle.

12. State that oestrogen stimulates the repair and vascularisation of the endometrium, thickening it and preparing it for implantation. It also affects the consistency of the cervical mucus, making it more easily penetrated by sperm.

13. State that peak levels of oestrogen stimulate a surge in the secretion of LH by the pituitary gland.

14. State that in the luteal phase (second half of the cycle), a surge in LH triggers ovulation.

15. State that ovulation is the release of an egg (ovum) from a follicle in the ovary.

16. State that ovulation usually occurs around the midpoint of the menstrual cycle.

17. State that LH also stimulates the development of the corpus luteum from the follicle and stimulates the corpus luteum to secrete the sex hormone progesterone.

18. State that progesterone promotes further development and vascularisation of the endometrium, preparing it for implantation if fertilisation occurs.

19. State that high levels of oestrogen and progesterone inhibit the secretion of FSH and LH by the pituitary gland, which prevents further follicles from developing.

20. State that the inhibition of FSH and LH by high levels of oestrogen and progesterone is an example of negative feedback control.

21. State that the lack of LH leads to degeneration of the corpus luteum with a subsequent drop in progesterone levels leading to menstruation.

22. State that if fertilisation does occur the corpus luteum does not degenerate and progesterone levels remain high.

Skills

1. *Process information to calculate a ratio.*
2. *Process information to calculate a percentage increase.*
3. *Select information from a line graph.*

2.3 The biology of controlling fertility

You should already know:

- The basic structure of sperm and egg cells.

Learning intentions

- Describe cyclical fertility in women and continuous fertility in men.
- Describe methods used to estimate the fertile period of a woman.
- Describe treatments for infertility in men and women.
- Describe physical and chemical methods of contraception.

Fertility

Fertility is the natural capability to produce offspring, which is dependent on age, health and other factors as shown in **Figure 2.3.1**.

Figure 2.3.1 *Factors affecting human fertility*

> 📖 **Contraception**
>
> The deliberate use of artificial methods or other techniques to prevent pregnancy.

Infertility treatments and **contraception** are based on the biology of fertility and the reproductive process.

Female fertility

Women show cyclical fertility leading to a fertile period. They are only fertile and able to conceive (become pregnant) for a few days during each menstrual cycle. The fertile period lasts for approximately five days around the time of ovulation.

Identification of the fertile period can be achieved from data on the timing of menstruation based upon ovulation occurring 14 days later, body temperature, **cervical mucus** viscosity and the life span of sperm and eggs.

A woman's body temperature rises by around 0·5°C after ovulation and her cervical mucus becomes thin and watery to allow the sperm to pass through into the uterus. Sperm are able to survive for several days after intercourse. This means that they can potentially still fertilise an ovum if ovulation takes place after intercourse while the sperm are still present and viable. **Figure 2.3.2** shows the fertile period and changes in temperature and cervical mucus viscosity in the menstrual cycle.

📖 Cervical mucus

Fluid secreted from the cervix which can vary in consistency from thin and watery to thick and sticky.

⚛ Make the link

There is more about the menstrual cycle in Chapter 2.2 on page 109.

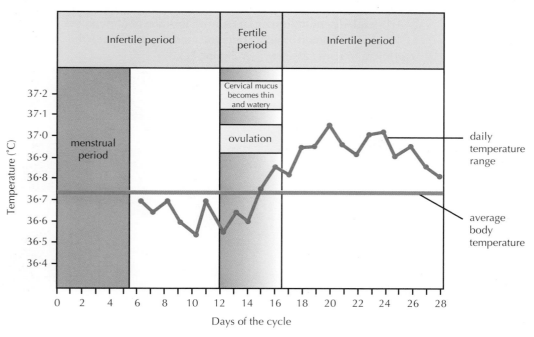

Figure 2.3.2 *The fertile period and changes in body temperature and cervical mucus viscosity in the menstrual cycle*

Male fertility

Men show **continuous fertility**. This is achieved through negative feedback control. Levels of FSH that stimulates sperm production and ICSH that stimulates testosterone production are maintained at a relatively constant level. This gives men the ability to continually produce sperm in their testes from puberty onwards.

⚛ Make the link

There is more about negative feedback control of testosterone and sperm production in Chapter 2.2 on page 108.

Treatments for infertility

Make the link

There is more about the roles of FSH, LH and negative feedback control in the menstrual cycle in Chapter 2.2 on page 110.

In vitro fertilisation (IVF)

Fertility treatment in a laboratory dish. *In vitro* means outside the body.

Stimulating ovulation

One form of female infertility is caused by failure to ovulate. This happens when the pituitary gland fails to secrete FSH to stimulate the development of a follicle in the ovary or LH which is needed to trigger ovulation. This form of infertility can be treated using drugs that act in different ways. Ovulation can be stimulated using **ovulatory drugs** that mimic the action of FSH and LH to stimulate egg production and ovulation. These drugs can cause super ovulation that can result in multiple births. They can also be used to collect ova for *in vitro fertilisation (IVF)* programmes (see below).

Other drugs can be used which prevent the negative feedback effect of oestrogen on the secretion of FSH. This increases FSH levels and stimulates the development of follicles in the ovary.

Artificial insemination

Artificial insemination is a type of fertility treatment that involves samples of semen being collected and injected into the uterus by artificial means, as shown in **Figure 2.3.3**.

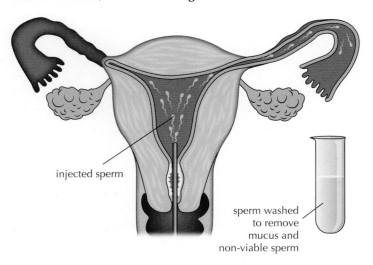

injected sperm

sperm washed to remove mucus and non-viable sperm

Figure 2.3.3 *Artificial insemination procedure*

Artificial insemination is particularly useful where the male has a low sperm count. If a male partner is sterile a donor may be used to provide semen.

In vitro fertilisation (IVF)

IVF is a type of fertility treatment where fertilisation takes place outside the body. It is suitable for people with a wide range of fertility issues and is one of the most commonly used and successful treatments available for many people. In the IVF procedure, the woman is treated using ovulatory drugs (fertility hormones) to stimulate the ovaries to produce several eggs. The eggs are then surgically removed, collected and mixed with sperm in a culture dish in a laboratory as shown in **Figure 2.3.4**. IVF is carried out when the sperm quality is considered to be 'normal'. If there are issues with the sperm quality such as low mobility or low sperm count, **intra-cytoplasmic sperm injection (ICSI)** may be used instead. If fertilisation is successful, the fertilised eggs are incubated and the embryos are allowed to develop for between two and six days until they have formed at least eight cells. **Pre-implantation genetic diagnosis (PGD)** is used to identify single gene disorders and chromosomal abnormalities. PGD can be offered to people who have a serious inherited condition in their family in order to avoid passing it to their children. The embryologist then selects the most suitable embryo, which is transferred back to the prepared uterus in the hope of successful implantation and birth.

It is normally best practice to freeze the remaining embryos because putting two embryos back in the uterus increases the chance of having twins or triplets, which carries health risks.

📖 **Intra-cytoplasmic sperm injection (ICSI)**

Fertility treatment that involves the injection of a sperm directly into an egg to achieve fertilisation.

📖 **Pre-implantation genetic diagnosis (PGD)**

Genetic profiling of embryos used to identify single gene disorders and chromosomal abnormalities.

Make the link

There is more about screening for genetic conditions in Chapter 2.4 on page 131.

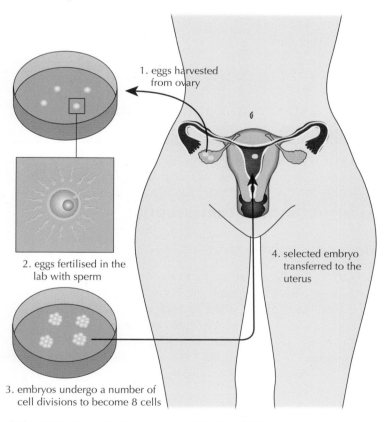

1. eggs harvested from ovary

2. eggs fertilised in the lab with sperm

3. embryos undergo a number of cell divisions to become 8 cells

4. selected embryo transferred to the uterus

Figure 2.3.4 *Stages in the* in vitro fertilisation *(IVF) procedure*

Intra-cytoplasmic sperm injection (ICSI)

For about half of couples who have problems conceiving, the cause of infertility is sperm-related. ICSI is the most common and successful treatment for male infertility.

ICSI can be used if mature sperm are defective (abnormally shaped or have poor mobility) or very low in number. The head of the sperm is detached from the tail and then the head only is drawn into a needle and injected directly into the egg to achieve fertilisation, as shown in **Figure 2.3.5**. ICSI itself is very successful at helping the sperm and the egg to fertilise – fertilisation happens in around 90% of cases. However, as in IVF, there are still many other factors affecting a successful pregnancy, including the age of the woman and whether she has any fertility difficulties herself.

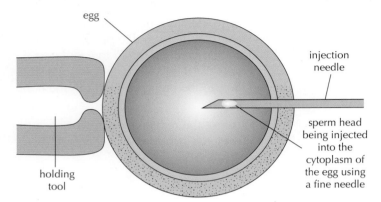

Figure 2.3.5 *The ICSI procedure used when sperm are defective or very low in number*

Physical and chemical methods of contraception

Contraception is the intentional prevention of pregnancy by natural or artificial methods, including both physical and chemical methods.

Physical methods of contraception

Barrier contraceptives

Barrier contraceptives prevent sperm from entering a woman's uterus. They include the condom, diaphragm, cervical cap and contraceptive sponge. Some condoms contain spermicides (chemicals that kill sperm). Spermicides should be used with condoms and other barrier contraceptives that do not already contain them. As a contraceptive, spermicide may be used alone. However, the pregnancy rate experienced by couples using only spermicide is higher than that of couples using other methods.

Intrauterine device (IUD)

An **intrauterine device (IUD)** is a small T-shaped plastic and copper device that is inserted into the uterus by a specially trained doctor or nurse as shown in **Figure 2.3.6**. The IUD works by stopping the sperm and egg from surviving in the womb or oviducts. It may also prevent a fertilised egg from implanting in the endometrium of the uterus. The IUD is a long-acting reversible contraceptive method.

The IUD releases copper, which changes the make-up of the fluids in the uterus and oviducts to stop sperm surviving there. IUDs may also stop fertilised eggs from implanting in the uterus. There are several different sizes and types of IUD, some with more copper than others. IUDs with more copper are more than 99% effective. This means that fewer than one in 100 women who use an IUD will get pregnant in one year. An IUD lasts for five to 10 years, depending on the type.

(a) **(b)**

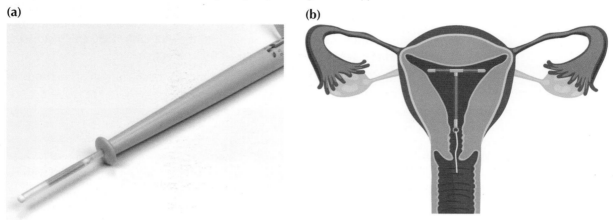

Figure 2.3.6 *(a) IUD applicator (b) IUD positioned in the uterus*

Sterilisation

Sterilisation in males is called a vasectomy. Vasectomy is a minor operation that is usually carried out under local anaesthetic and takes about 15 minutes.

The tubes (sperm ducts) that carry sperm from the testes to the penis are cut, blocked or sealed as shown in **Figure 2.3.7(a)**. This prevents sperm from reaching the seminal fluid (semen), which is ejaculated from the penis during intercourse. There will be no sperm in the semen, so an egg cannot be fertilised. In most cases, vasectomy is more than 99% effective.

> ### Make the link
>
> An IUS (intrauterine system) is like an IUD but can release the hormone progesterone and so is both a physical and chemical method of contraception.

(a) **(b)**

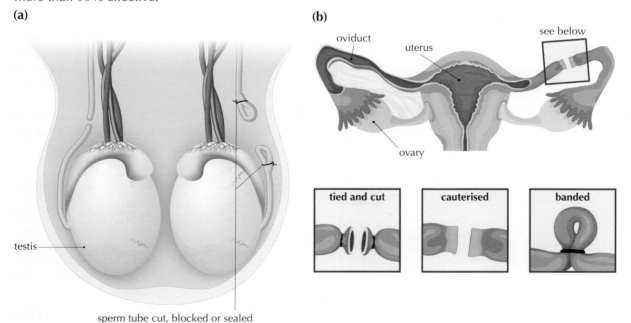

Figure 2.3.7 *(a) Sterilisation in males (b) Sterilisation in females*

Female sterilisation is an operation carried out under general anaesthetic that blocks the oviducts which link the ovaries to the uterus. This can be done by applying plastic or titanium clips or by tying and cutting the oviducts as shown in **Figure 2.3.7(b)**. This prevents the woman's eggs from reaching sperm and becoming fertilised. Eggs will still be released from the ovaries as normal, but they will be absorbed naturally into the woman's body. In most cases, female sterilisation is more than 99% effective.

The following table provides a summary of the physical methods of contraception.

Physical methods of contraception	
Method	**Description**
Condom	Prevents sperm entering the female and can be used with spermicidal gel to kill sperm
Cervical cap/diaphragm	Prevents sperm entering the uterus and can be used with spermicidal gel
Intrauterine device (IUD)	Small copper and plastic device inserted into the uterus to prevent implantation
Sterilisation (vasectomy and tubal ligation)	Surgical procedures to prevent sperm reaching the egg by cutting the sperm ducts or oviducts

Chemical methods of contraception

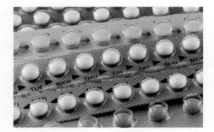

Figure 2.3.8 *Contraceptive pills*

The oral contraceptive pill is a chemical method of contraception. The **combined contraceptive pill** contains a combination of synthetic oestrogen and progesterone that mimics negative feedback to prevent the release of FSH and LH from the pituitary gland. The hormones in the pill prevent the ovaries from releasing an egg (ovulating). They also make it difficult for sperm to reach an egg by thickening the cervical mucus, or for an egg to implant itself by thinning the lining of the uterus. When taken correctly, the pill is over 99% effective at preventing pregnancy. However, due to human error it is considered to be 92% effective.

The **progesterone-only pill (POP)** or **mini pill** stops sperm reaching an egg by causing thickening of the cervical mucus.

The emergency contraceptive pill, sometimes called the **morning-after pill**, contains a high dose of synthetic progesterone and oestrogen and works by preventing ovulation or implantation.

The following table provides a summary of the chemical methods of contraception.

Chemical methods of contraception	
Method	**Description**
Combined contraceptive pill	Contains a combination of synthetic oestrogen and progesterone that mimics negative feedback to prevent the release of FSH and LH from the pituitary gland
Progesterone-only pill (POP)/mini pill	Causes thickening of the cervical mucus and so prevents sperm entering the uterus
Morning-after pill (emergency contraceptive pill)	Contains a high dose of synthetic oestrogen and progesterone to prevent ovulation or implantation

GO! Activity 2.3.1 Work individually to ...

Structured questions

1. Infertility treatments and contraception are based on the biology of fertility and the reproductive process.

    a) Artificial insemination and intra-cytoplasmic sperm injection (ICSI) are fertility treatments that may be used if a man has a low sperm count.

        Describe each of these treatments.   2

    b) **Explain** the role of ovulatory drugs in the treatment of female infertility.   2

    c) **Explain** the mechanism of action of the combined contraceptive pill.   2

2. Data from *in vitro* fertilisation (IVF) clinics is used to indicate how a woman's age can affect the success rate of IVF.

    a) The graph shows the effect of age on the percentage success of having a baby after using IVF in one clinic. The data refers to women using their own eggs (rather than from a donor).

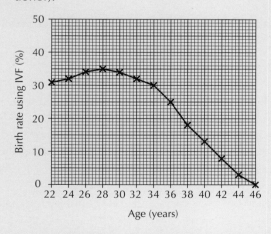

        i. **Use values from the graph to describe** how the IVF birth rate changes with the age of the women treated at this clinic.   2

(Continued)

ii. **Calculate** how many 28-year-old women in a sample of 800 would give birth following treatment with IVF at this clinic. 1

b) Some women are given the option to use eggs from a donor when undergoing IVF. This can increase their chances of becoming pregnant and giving birth.

The table shows the birth rates for women of different ages after undergoing IVF using donor eggs.

Age (years)	22	26	30	34	38	42	46
Birth rate (%)	45	65	60	57	54	48	39

i. Using data in the table and graph, **calculate** the difference in birth rate for women aged 38 when using donor eggs rather than their own eggs. 1

ii. **Calculate** the ratio of women aged 42 giving birth due to IVF treatment using donor eggs compared to those using their own eggs. 1

c) Some women undergoing IVF consent to pre-implantation genetic diagnosis (PGD) of their embryos.

Explain why PGD is offered to some women. 1

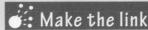 **Make the link**

There is information about selecting information and processing information in Chapter 4 on page 267.

Extended response questions

1. **Explain** the biological basis for the use of ovulatory drugs in fertility treatment. 5
2. **Discuss** the biological basis of contraception. 9

 Activity 2.3.2 Work in pairs to …

1. Flashcard activity

You will each need: a set of blank flashcards (A7 cards) and a stopwatch.

- Find the glossary terms for this chapter – they are the **black** typeface and **red** typeface terms. Using your blank cards, you should each make a set of flashcards for these terms – write the term on one side and the definition on the other. You will find the definitions in the chapter.

- Shuffle your cards and lay them out in a column, some showing terms and some showing definitions – you decide. Your partner should match their cards with yours, laying their cards in a column beside yours to give the corresponding term or definition. Time how long they take to do this.

- Now swap roles – your partner should lay out their cards and you should try the matching exercise while your partner times you.

- You should each keep your set of flashcards as a revision tool for later.

Activity 2.3.3 Work as a group to ...

1. **Research the causes of infertility.**

   Visit the NHS web page below:

   www.nhs.uk/conditions/infertility/

   Split your group into four to research the following sub-topics:

   - Polycystic ovary syndrome (PCOS)
   - Endometriosis
   - Pelvic inflammatory disease (PID)
   - Male infertility

   Each group should find information on symptoms, causes and treatments as appropriate and produce a two- or three-slide PowerPoint presentation or storyboard to summarise their research.

   Combine the slides or storyboards to form a presentation or display and be prepared to present it to your class.

2. **Placemat activity: Contraception**

   You will need: a placemat template (Appendix 3) and four fine marker pens.

   - Set the placemat in the middle of the table and each write your name in a section.
   - Each participant should then write words that they think are related to *contraception* into their section of the placemat. Spend 2 minutes doing this.
   - You should take it in turns to read out a word from your section. If everyone agrees it is related to the topic it can be copied into the centre section of the placemat. Continue until all words have been discussed.
   - Working as a group, use all the words in the centre section to summarise your knowledge of the physical and chemical methods of contraception.

3. **Placemat activity: Fertility treatments**

   You will need: a placemat template (Appendix 3) and four fine marker pens.

   - Set the placemat in the middle of the table and each write your name in a section.
   - Each participant should then write words that they think are related to *fertility treatments* into their section of the placemat. Spend 2 minutes doing this.
   - You should take it in turns to read out a word from your section. If everyone agrees it is related to the topic it can be copied into the centre section of the placemat. Continue until all words have been discussed.
   - Working as a group, use all the words in the centre section to summarise your knowledge of fertility treatments.

Learning checklist

After working on this chapter, I can:

Knowledge and understanding

1. State that infertility treatments and contraception are based on the biology of fertility.

2. State that men continually produce sperm in their testes so show continuous fertility.

3. State that continuous fertility in men is due to the relatively constant levels of pituitary hormones.

4. State that women are only fertile for a few days during each menstrual cycle.

5. State that women show cyclical fertility leading to a fertile period.

6. State that the time with the highest likelihood of pregnancy resulting from sexual intercourse is from a few days before until 1-2 days after ovulation.

7. State that the time of ovulation can be estimated based upon a slight rise in body temperature on the day of ovulation and the thinning of cervical mucus.

8. State that a woman's body temperature rises by around 0·5°C after ovulation and her cervical mucus becomes thin and watery.

9. State that female infertility may be due to failure to ovulate, which is usually the result of a hormone imbalance.

10. State that ovulatory or fertility drugs can be used to stimulate ovulation.

11. State that some ovulatory drugs work by preventing the negative feedback effect of oestrogen on FSH secretion.

12. State that other ovulatory drugs mimic the action of FSH and LH.

13. State that ovulatory drugs can cause super ovulation that can result in multiple births or can be used to collect ova for *in vitro* fertilisation (IVF) programmes.

14. State that artificial insemination is a treatment in which semen is inserted into the female reproductive tract without intercourse having taken place.

15. State that artificial insemination is particularly useful where the male has a low sperm count.

16. State that if a male partner is infertile (sterile), a donor may be used to provide semen for artificial insemination.

17. State that IVF involves the surgical removal of eggs from ovaries after hormone stimulation, mixing with sperm in a culture dish to achieve fertilisation, incubation of zygotes and uterine implantation.

18. State that IVF can be used in conjunction with pre-implantation genetic diagnosis (PGD) to identify single gene disorders and chromosomal abnormalities.

19. State that eggs fertilised by IVF are incubated until they have formed at least eight cells and are then transferred to the uterus for implantation.

20. State that if mature sperm are defective or very low in number, intra-cytoplasmic sperm injection (ICSI) can be used.

21. State that ICSI involves the head of a sperm being drawn into a needle and injected directly into the egg to achieve fertilisation.

22. State that contraception is the intentional prevention of pregnancy (conception) by natural or artificial methods and includes both physical and chemical methods.

23. State that physical methods of contraception have a biological basis and include barrier methods, intrauterine devices (IUD) and sterilisation procedures.

24. State that the oral contraceptive pill is a chemical method of contraception.

25. State that the oral contraceptive pill contains a combination of synthetic oestrogen and progesterone that mimics negative feedback to prevent the release of FSH and LH from the pituitary gland.

26. State that the progesterone-only pill (POP) (mini pill) causes thickening of the cervical mucus to prevent the entry of sperm and fertilisation.

27. State that the morning-after pill prevents ovulation or implantation.

Skills

1. *Process information to calculate an arithmetical difference and ratios.*

2. *Select information from a line graph.*

2.4 Antenatal and postnatal screening

You should already know:

- The genetic terms: gene; allele; phenotype; genotype; dominant; recessive; homozygous; heterozygous and P, F_1 and F_2.
- The method of carrying out a monohybrid cross from a parental generation through to the F_2 generation, including the use of Punnett squares.
- The reasons why predicted phenotype ratios among offspring are not always achieved.
- The method of identifying the phenotypes and genotypes of members of a family tree.

Learning intentions

- Describe techniques used to monitor the health of the mother, developing fetus and baby: antenatal screening including ultrasound imaging, blood and urine tests and diagnostic tests including amniocentesis and chorionic villus sampling (CVS).
- Describe and analyse patterns of inheritance in genetic disorders, interpreting family histories over three generations.
- Describe patterns of inheritance in autosomal recessive, autosomal dominant, incomplete dominance and sex-linked recessive single gene disorders.
- Describe the diagnostic testing of phenylketonuria (PKU), explain how it is caused and how it is treated.

📖 Antenatal screening

Tests carried out before birth to find out if the mother or baby are at risk of having a disorder.

📖 Ultrasound scan

Diagnostic procedure used for various prenatal checks, such as establishing the stage of pregnancy and the due date or detecting physical abnormalities.

📖 Marker chemicals

Chemicals produced during pregnancy and tested for in conjunction with scans to test for various conditions.

Antenatal screening

A variety of techniques can be used to monitor the health of the mother, developing fetus and baby.

Antenatal screening identifies the risk of a disorder so that further tests and a prenatal diagnosis can be offered.

Ultrasound imaging

Pregnant women are given two **ultrasound scans**. **Dating scans** which determine pregnancy stage and due date are used with tests for **marker chemicals** which vary normally during pregnancy. **Anomaly scans** may detect serious physical abnormalities in the fetus. A dating scan takes place between 8 and 14 weeks and an anomaly scan between 18 and 20 weeks.

(a)  (b)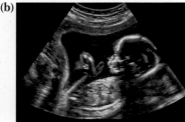

Figure 2.4.1 *(a) Ultrasound scan (b) Ultrasound image of the fetus*

Blood and urine tests

A mother's blood transports a range of chemicals round the body, including the products of digestion, hormones and substances that have crossed the placenta from the fetus. Some of these chemicals show significant changes if there are problems with either the fetus or the mother, and can be used as chemical markers.

Routine blood and urine tests are carried out throughout pregnancy to monitor the concentrations of marker chemicals and compared with those expected at that stage in the pregnancy as shown in **Figure 2.4.2**. Measuring a chemical at the wrong time or stage of pregnancy could lead to a **false positive result**.

An atypical chemical concentration can lead to **diagnostic testing** being carried out to determine if the fetus has a medical condition.

Diagnostic testing

Amniocentesis and **chorionic villus sampling (CVS)** are diagnostic tests used to confirm chromosomal conditions such as Down syndrome and fetal infections.

Amniocentesis is a prenatal test in which a small amount of amniotic fluid is removed from the sac surrounding the fetus for testing. The sample of amniotic fluid is removed through a fine needle inserted into the uterus through the abdomen, under ultrasound guidance as shown in **Figure 2.4.3**. Amniocentesis presents a small risk of miscarriage. This prenatal test is generally offered to women who have a significant risk for genetic diseases, including those who have an abnormal ultrasound or abnormal lab screens, have a family history of certain birth defects or have previously had a pregnancy or child with a birth defect.

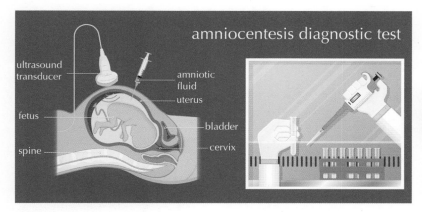

amniocentesis diagnostic test

ultrasound transducer
amniotic fluid
uterus
fetus
bladder
spine
cervix

Figure 2.4.3 *Amniocentesis*

Chorionic villus sampling is also a prenatal diagnostic test to detect birth defects, genetic diseases and other problems during pregnancy. Chorionic villi are tiny parts of the placenta that are formed from the fertilised egg through which exchange of materials takes place during pregnancy.

📖 **False positive result**

When the test result indicates the presence of a condition that is absent.

📖 **Diagnostic testing**

Tests such as amniocentesis and CVS that are used to confirm if the fetus has a medical condition.

Figure 2.4.2 *Routine blood test to monitor marker chemicals during pregnancy*

📖 **Amniocentesis**

A prenatal diagnostic test used to confirm chromosomal conditions such as Down syndrome.

📖 **Chorionic villus sampling (CVS)**

A prenatal diagnostic test that samples cells from the placenta to detect birth defects and chromosomal conditions such as Down syndrome.

During the test, a small sample of cells from the chorionic villi is taken from the placenta as shown in **Figure 2.4.4**.

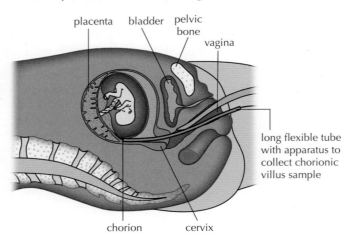

Figure 2.4.4 *Chorionic villus sampling procedure*

There is more about homologous chromosomes in Chapter 1.1 on page 12 and Chapter 1.4 on page 47.

📖 Karyotype

Display of chromosomes arranged as homologous pairs produced for medical purposes.

📖 Homologous pair

A matching pair of chromosomes, one from the mother and one from the father, which have the same genes at the same position.

⚗ Make the link

CVS can be carried out earlier in pregnancy than amniocentesis, although it has a higher risk of miscarriage. Cells from samples can be cultured to obtain sufficient cells to produce a **karyotype** to diagnose a range of conditions.

A karyotype shows an individual's chromosomes arranged as **homologous pairs**. Any abnormalities in terms of the numbers or shapes of the chromosomes can help diagnose genetic conditions such as Down syndrome as shown in **Figure 2.4.5**.

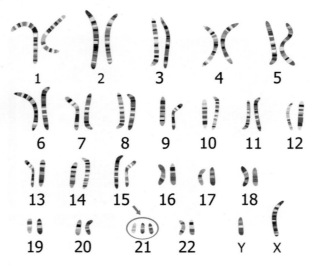

Figure 2.4.5 *Karyotype showing the chromosome complement of a boy with Down syndrome*

The element of risk will be assessed when deciding to proceed with these tests, as will any decisions the individuals concerned are likely to make if a test is positive.

Analysis of patterns of inheritance in genetic screening and counselling

Genetic screening can be carried out to determine whether a person is carrying a genetic mutation that causes a particular medical condition. A pattern of inheritance can be revealed by collecting information about a particular characteristic from family members and using it to construct a family tree. If the phenotypes of the individuals are known, most of the genotypes of the individuals can also be determined. The family tree provides information which a genetic counsellor can use to advise parents in situations where there is the possibility of passing on a genetic disorder to potential offspring. It can be used to calculate the percentage chance of inheriting a single gene disorder such as cystic fibrosis, thalassemia, Huntington's disease and haemophilia.

Autosomal recessive gene disorders

An **autosome** is a chromosome that is not a sex chromosome. Diploid cells in humans contain 22 pairs of autosomes and one pair of sex chromosomes (XX in females and XY in males).

An **autosomal recessive** disorder, such as cystic fibrosis (CF), is expressed relatively rarely in the offspring. It affects males and females equally and may skip generations. Cystic fibrosis is an inherited condition which causes the production of thick mucus in the respiratory system. It also affects the pancreas, liver, kidneys and intestine. **Figure 2.4.6** shows the effect of cystic fibrosis on the lungs and pancreas.

> ### 📖 Autosomal recessive
>
> An allele on chromosomes 1–22 (the autosomes) only expressed in the phenotype if the genotype is homozygous for the recessive allele.

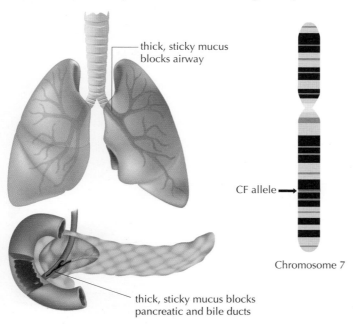

thick, sticky mucus blocks airway

CF allele →

Chromosome 7

thick, sticky mucus blocks pancreatic and bile ducts

Figure 2.4.6 *Effect of cystic fibrosis on the lungs and pancreas and the location of the allele on chromosome 7*

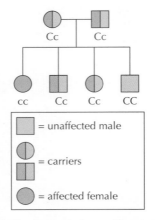

Figure 2.4.7 *Cystic fibrosis in a family tree*

📖 Autosomal dominant

An allele on chromosomes 1–22 (the autosomes) which is always expressed in the phenotype whether the genotype is homozygous or heterozygous.

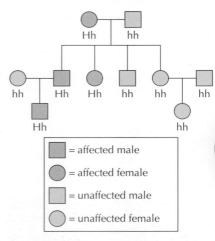

Figure 2.4.8 *Huntington's disease in a family tree*

📖 Autosomal incomplete dominant

When an allele is not completely masked by a dominant allele and so affects an individual's phenotype.

One in 25 of us carry the recessive CF allele. Long-term issues include difficulty breathing and coughing up mucus as a result of frequent lung infections. The inheritance of cystic fibrosis in a family is shown in **Figure 2.4.7**. Hereditary carriers of the condition have inherited a recessive allele and do not show the condition but can pass it to their offspring.

Autosomal dominant gene disorders

Huntington's disease is an inherited degenerative disorder that results in the death of brain cells. It is characterised by slurred speech, uncontrolled movements of the body and a progressive deterioration in mental functions. The first symptoms generally appear between the ages of 30 and 50. The condition becomes progressively worse and is usually fatal after a period of up to 20 years.

An **autosomal dominant** condition such as Huntington's disease shows up in every generation and affects males and females equally as shown in **Figure 2.4.8**.

Autosomal incomplete dominance gene disorders

In examples of **autosomal incomplete dominant** gene disorders, the fully expressed form of the condition is rare, the partly expressed form is more common and males and females are affected equally.

Sickle-cell disease is the name for a group of inherited health conditions that affect the red blood cells. Sickle-cell disease is particularly common in people with an African or Caribbean family background. People with sickle-cell disease produce unusually shaped red blood cells that can cause problems because they do not live as long as healthy blood cells and can block blood vessels as shown in **Figure 2.4.9**.

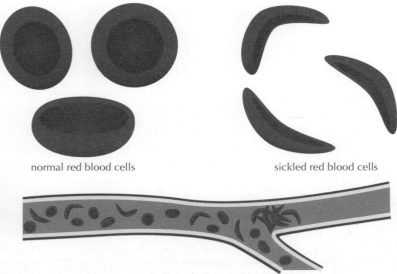

normal red blood cells   sickled red blood cells

Figure 2.4.9 *Sickle blood cells disrupting the blood flow in a blood vessel*

The inheritance of sickle-cell disease in a family is shown in **Figure 2.4.10**.

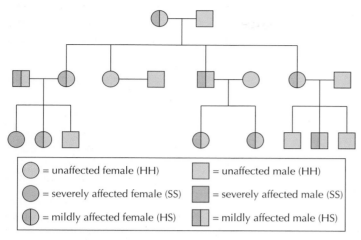

Figure 2.4.10 *Sickle-cell disease in a family tree*

Carriers of the allele have a less severe form of the disease called sickle-cell trait, while those with two of the alleles have the more severe condition called sickle-cell anaemia. Since neither allele is completely dominant, each allele is represented by a different upper case letter. H is used to represent the normal allele, while S represents the allele for haemoglobin S.

- HH individuals are normal.
- HS individuals have sickle-cell trait – the red blood cells are generally normal but individuals may show some signs of the disease when carrying out intense physical activity. This occurs because the H allele is not completely dominant to the S allele, which means that S is partially expressed.
- SS individuals have sickle-cell anaemia – the red blood cells stick together, causing problems in the circulatory system, which can lead to severe organ damage and usually death.

Sex-linked recessive gene disorders

In **sex-linked recessive** disorders, males are affected more than females. The allele is carried on the X-chromosome, of which males have only a single copy. Male offspring inherit the condition from their mother. Fathers cannot pass the condition on to their sons and female offspring can only be affected if the father has the condition and the mother is at least a carrier.

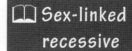

📖 Sex-linked recessive

A recessive allele carried on the X-chromosome.

Haemophilia is a sex-linked recessive condition in which a protein, clotting factor VIII, is defective. Males who suffer from the disease carry a mutated allele of the factor VIII gene, which means that their blood fails to clot properly. Heterozygous females are carriers and possess one mutated allele, but do not suffer from haemophilia. There is a 50% chance that any daughter of a female carrier will also be a carrier and a 50% chance that a son will have the disorder, assuming

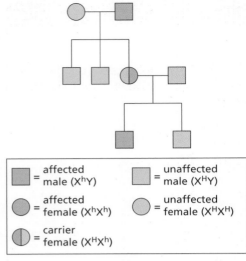

■ = affected male (X^hY)	■ = unaffected male (X^HY)	
● = affected female (X^hX^h)	● = unaffected female (X^HX^H)	
◐ = carrier female (X^HX^h)		

Figure 2.4.11 *Haemophilia in a family tree*

📖 Phenylketonuria (PKU)

A rare genetic condition resulting in a metabolic disorder that is tested for by postnatal screening.

⚗ Make the link

There is more about metabolic pathways in Chapter 1.6 on page 60.

that the father is not a haemophiliac himself. Any daughter of a male haemophiliac will be a carrier and all of the sons will be unaffected because they inherit the X-chromosome from their mother, assuming that she is not a carrier.

Red–green colour blindness in humans is also a sex-linked recessive gene disorder.

Figure 2.4.11 shows inheritance of the sex-linked recessive condition haemophilia in a family.

Postnatal screening

Postnatal screening involves health checks that are carried out after the birth of a baby. These are aimed at detecting certain conditions or abnormalities. Postnatal diagnostic testing is used to detect metabolic disorders such as **phenylketonuria (PKU)**.

PKU is a rare genetic condition that affects an individual's metabolism.

In PKU, a substitution mutation means that the enzyme that converts the amino acid phenylalanine to tyrosine is non-functional. About 1 in 10,000 babies born in the UK has PKU. Without treatment, the babies brain and nervous system can be damaged which can lead to learning disabilities and behavioural difficulties such as frequent temper tantrums.

Newborn blood spot screening is carried out on babies at around 5 days old. This is done to test for PKU and many other conditions. This involves pricking the baby's heel to collect drops of blood to test as shown in Figure 2.4.12.

Figure 2.4.12 *Nurse performing the neonatal heel-prick test for PKU*

If PKU is confirmed, treatment is given immediately to reduce the risk of serious complications. Treatment includes a special protein restricted diet low in phenylalanine that completely avoids high protein foods such as meat, eggs, fish and dairy products. Individuals diagnosed with PKU must take an amino acid supplement to ensure they are getting all the nutrients required for normal growth and good health and are monitored through regular blood tests. Early diagnosis and treatment means that most children with PKU are able to live healthy lives.

GO! Activity 2.4.1 Work individually to ...

Structured questions

1. **a)** Chorionic villus sampling (CVS) is a prenatal diagnostic test that samples cells from the placenta to detect birth defects and chromosomal conditions such as Down syndrome. The cells obtained from CVS are used to prepare a karyotype.

   i. Explain the difference between a screening test and a diagnostic test such as CVS. 2

   ii. Describe the process by which a karyotype is produced. 2

   iii. Discuss the advantages and disadvantages of using CVS rather than amniocentesis during antenatal screening. 2

   b) Name the type of antenatal screening tests which are routinely carried out to monitor the concentration of certain substances, such as protein, in a pregnant woman's blood. 1

2. Haemophilia is a sex-linked disorder that results in an individual producing a faulty blood-clotting protein.

   The diagram below shows the inheritance of haemophilia in a family.

   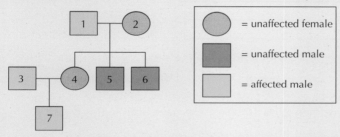

   a) The condition is caused by a recessive sex-linked allele represented by the letter h.

   i. State the genotypes of individuals 3 and 4. 2

   ii. Explain why individual 1 could not pass the condition to his sons. 1

   iii. Individual 6 has a son with a woman who is a carrier of the condition.

   Calculate the percentage chance of their son having this condition. 1

   b) The condition occurs with a frequency of 1 in 350 males.

   Assuming an equal proportion of males and females, **calculate** how many males are likely to have the condition in a town with a population of 175,000. 1

   c) Huntington's disease is caused by a single dominant allele, which is not sex-linked.

   A woman's father is heterozygous for this condition and her mother is unaffected.

   Calculate the chances that this woman may have inherited the condition. 1

3. **a)** Phenylketonuria (PKU) is an example of a genetic disorder which affects the following metabolic pathway.

   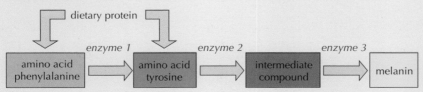

   Babies born with PKU can develop brain damage from the build-up of phenylalanine and its harmful metabolites.

(Continued)

    i. All babies are tested for PKU immediately after birth.

        State the term used to describe this type of diagnostic testing.   1

    ii. Describe how brain damage can be prevented in babies diagnosed with PKU.   1

  b) PKU is caused by an autosomal recessive allele.

    A couple, who are both unaffected, have a child who has PKU.

    Calculate the percentage chance of their next child having this disorder.   1

4. The table shows how the risk of Down syndrome in live births varies with maternal age in a European country.

Maternal age (years)	Risk of Down Syndrome (% of live births)
20	0·1
25	0·1
30	0·2
35	0·5
40	0·8
45	3·6

  a) On a separate piece of graph paper, **plot a line graph** to show the data in the table.   2

  b) Calculate the percentage increase in risk of Down syndrome as maternal age rises from 30 to 45 years.   1

  c) In one year, the number of women aged 40 years who gave birth in this country was 500.

    Calculate the number of babies expected to have Down syndrome.   1

  d) Based on the data, **predict** the % risk of Down syndrome in a baby born to a 27 year old woman.   1

Extended response questions

1. Explain the use of family tree charts in genetic screening and counselling.   4

2. Discuss the screening and testing procedures which may be carried out as part of antenatal care.   9

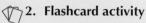 **Activity 2.4.2 Work in pairs to …**

1. **Placemat activity: Antenatal screening techniques**

   You will need: a placemat template (Appendix 3) and four fine marker pens.

   - Set the placemat in the middle of the table and each write your name in a section.
   - Each participant should then write words that they think are related to *techniques used in antenatal screening* into their section of the placemat. Spend 2 minutes doing this.
   - You should take it in turns to read out a word from your section. If everyone agrees it is related to the topic it can be copied into the centre section of the placemat. Continue until all words have been discussed.
   - Working as a group, use all the words in the centre section to summarise your knowledge of techniques used in antenatal screening.

2. **Flashcard activity**

   You will each need: a set of blank flashcards (A7 cards) and a stopwatch.

   - Find the glossary terms for this chapter – they are the **black** typeface and **red** typeface terms. Using your blank cards, you should each make a set of flashcards for these terms – write the term on one side and the definition on the other. You will find the definitions in the chapter.
   - Shuffle your cards and lay them out in a column, some showing terms and some showing definitions – you decide. Your partner should match their cards with yours, laying their cards in a column beside yours to give the corresponding term or definition. Time how long they take to do this.
   - Now swap roles – your partner should lay out their cards and you should try the matching exercise while your partner times you.
   - You should each keep your set of flashcards as a revision tool for later.

Activity 2.4.3 Work as a group to …

1. **Design and make a spider diagram to show the techniques used in antenatal and postnatal screening.**

   You will need: an A2 sheet and medium tip marker pens.

   Your teacher may ask your group to present the work to the class.

 2. **Research screening tests.**

   Visit the NHS web page below:

   www.nhs.uk/conditions/pregnancy-and-baby/screening-tests-in-pregnancy/

   Split your group into two to research the following sub-topics:

   - Antenatal screening – select two of the screening tests
   - Postnatal screening – select two of the screening tests

   Each group should produce a two-slide PowerPoint presentation to summarise their research and be prepared to present it to the class.

Learning checklist

After working on this chapter, I can:

Knowledge and understanding

1. State that a variety of techniques can be used to monitor the health of the mother, developing fetus and baby.

2. State that antenatal or prenatal screening involves testing for diseases or conditions in a fetus or embryo before it is born.

3. State that antenatal screening identifies the risk of a disorder so that further tests and a prenatal diagnosis can be offered.

4. State that common antenatal testing procedures include ultrasound scanning, amniocentesis and chorionic villus sampling (CVS).

5. State that an ultrasound scanner is used to produce an ultrasound image on a computer screen.

6. State that pregnant women are given two ultrasound scans.

7. State that dating scans which determine pregnancy stage and due date are used with tests for marker chemicals which vary normally during pregnancy.

8. State that a dating scan takes place between 8 and 14 weeks.

9. State that anomaly scans may detect serious physical abnormalities in the fetus.

10. State that an anomaly scan takes place between 18 and 20 weeks.

11. State that routine blood and urine tests are carried out throughout pregnancy to monitor the concentrations of marker chemicals.

12. State that marker chemicals are produced during normal physiological changes that take place during pregnancy.

13. State that measuring a chemical at the wrong time could lead to a false positive result.

14. State that an atypical chemical concentration can lead to diagnostic testing being carried out to determine if the fetus has a medical condition.

15. State that diagnostic tests include amniocentesis and chorionic villus sampling (CVS) from the placenta.

16. State that amniocentesis and CVS allow a prenatal diagnosis to be made and can confirm the presence of conditions such as Down syndrome.

17. State that an amniocentesis procedure has a small risk of miscarriage.

18. State that CVS can be carried out earlier in pregnancy than amniocentesis, but it has a higher risk of miscarriage.

19. State that cells from an amniocentesis sample or CVS can be cultured to obtain sufficient cells to produce a karyotype to diagnose a range of conditions.

20. State that a karyotype shows an individual's chromosomes arranged as homologous pairs.

21. State that in deciding to proceed with amniocentesis and CVS tests, the element of risk will be assessed, as will the decisions the individuals concerned are likely to make if a test is positive.

22. State that pedigree charts (family trees) are compiled and used in genetic screening and counselling to analyse patterns of inheritance.

23. State that pedigree charts are constructed to provide information and advice in situations where there is the possibility of passing on a genetic disorder to potential offspring.

24. State that pedigree charts can be used to analyse patterns of inheritance involving autosomal recessive, autosomal dominant, incomplete dominance and sex-linked recessive single gene disorders.

25. State that alleles are forms of the same gene.

26. State that homozygous individuals have two copies of the same allele and heterozygous individuals have copies of two different alleles of the same gene.

27. State that an autosomal recessive disorder, such as cystic fibrosis (CF), is expressed relatively rarely in the offspring. It affects males and females equally and may skip generations.

28. State that an autosomal dominant disorder such as Huntington's disease (HD) shows up in every generation and affects males and females equally.

29. State that in examples of autosomal incomplete dominance, the fully expressed form of the condition is rare, the partly expressed form is more common and males and females are affected equally.

30. State that in sex-linked recessive disorders, males are affected more than females. Male offspring receive the condition from their mother; fathers cannot pass the condition on to their sons and female offspring can only be affected if the father has the condition and the mother is at least a carrier.

31. State that postnatal screening involves health checks that are carried out after the birth of the baby and are aimed at detecting certain conditions or abnormalities.

32. State that postnatal diagnostic testing is used to detect metabolic disorders such as phenylketonuria (PKU).

33. State that PKU is an inborn error of metabolism caused by an autosomal recessive genetic disorder.

34. State that in PKU, a substitution mutation means that the enzyme which converts phenylalanine to tyrosine is non-functional.

35. State that if PKU is not detected soon after birth the baby's mental development can be affected.

36. State that individuals with high levels of phenylalanine are placed on a restricted diet that lacks the amino acid phenylalanine.

Skills

1. *Present information as a line graph.*

2. *Use tabular information to calculate a percentage increase.*

3. *Process information in general calculations.*

4. *Make a prediction from tabulated information.*

2.5 The structure and function of arteries, capillaries and veins

You should already know:

- Arteries have thick, muscular walls, a narrow central channel and carry blood under high pressure away from the heart.
- Veins have thinner muscular walls, a wider channel and carry blood under low pressure back towards the heart.
- Veins contain valves to prevent backflow of blood.
- Capillaries are thin walled and have a large surface area, forming networks at tissues and organs to allow efficient exchange of materials.
- Tissues contain capillary networks to allow the exchange of materials at cellular level.

Learning intentions

- Describe blood circulation to and from the heart through the arteries to the capillaries and then to the veins.
- Describe the structure and function of arteries, capillaries and veins.
- Explain changes in blood pressure as blood circulates.
- Explain how vasoconstriction and vasodilation can control blood flow.
- Explain the role of pressure filtration in exchange of material in capillaries.
- Describe the role of lymphatic vessels.

Circulation of blood

Blood is pumped from the **heart** every time it beats. As shown in **Figure 2.5.1(a)**, the heart occupies an almost central position in the chest, protected by the ribcage and muscles. The blood is circulated at high pressure from the heart through the **arteries**, which branch to visit all the areas of the body as outlined in **Figure 2.5.1(b)**. There is a decrease in blood pressure as it moves further away from the heart. Blood from small arteries flows into networks of **capillaries** with the body organs where exchange of material between the blood and the cells occurs. Small **veins** lead the blood away from the capillaries and into large veins for return to the heart at low pressure. **Figure 2.5.2** shows how the different types of blood vessel are connected to each other.

📖 Artery
Blood vessel that carries blood away from the heart.

📖 Capillary
Narrow, thin-walled blood vessel that exchanges materials with the tissues.

📖 Vein
Blood vessel with valves that transports blood back to the heart.

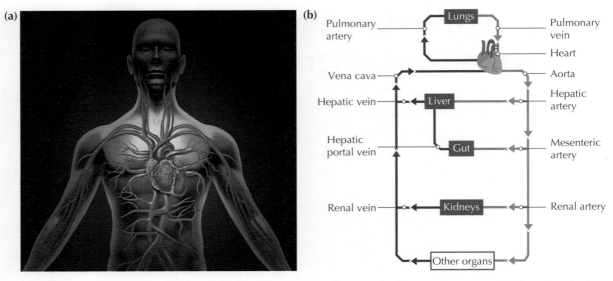

Figure 2.5.1 *(a) Image of part of the human circulatory system showing the heart and some arteries in red and veins in blue (b) Plan of circulation showing the heart and some arteries in red and veins in blue*

> ## 🔍 Hint
> You don't have to learn the names of arteries or veins apart from those which lead to and from the heart directly.

> ## 🔍 Hint
> Arteries carry blood Away from the heart.

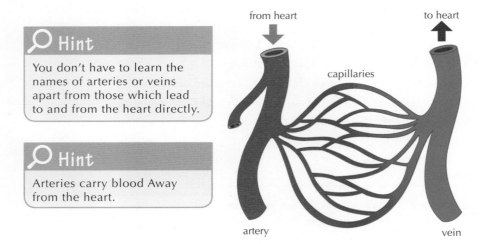

Figure 2.5.2 *Relationship between arteries, capillaries and veins*

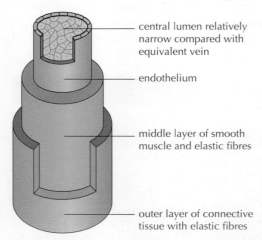

Figure 2.5.3 *Structure of an artery*

Blood vessels

Structure and function of arteries

Arteries have an outer layer of connective tissue containing **elastic fibres** and a middle layer containing **smooth muscle** with more elastic fibres as shown in **Figure 2.5.3**. Arteries have a relatively narrow **lumen**. The elastic walls of the arteries stretch and recoil to accommodate the surge of blood after each contraction of the heart. The recoil allows the blood pressure to be maintained as blood passes through the main arteries. The smooth muscle in the artery walls allows the diameter of the arteries to be narrowed and widened during **vasoconstriction** and **vasodilation**. Arteries have a layer of **endothelium** as their lining.

Structure and function of capillaries

Capillaries have a microscopically narrow lumen, only wide enough to let one red blood cell through at a time. The walls are made up of an endothelial layer one cell thick and each cell is loosely joined to others, leaving slits through which blood **plasma** can escape.

Capillaries branch to give complex networks, tens of thousands of kilometres long, in which the exchange of substances with tissues occurs through their super-thin walls as shown in **Figure 2.5.4(b)**. As blood flows through the capillary networks, blood pressure is reduced due to leakage of fluid from the capillary walls through slits between cells.

📖 Vasoconstriction
Narrowing of blood vessels to reduce blood flow.

📖 Vasodilation
Widening of blood vessels to increase blood flow.

📖 Endothelium
Layer of cells that lines the inner surface of blood vessels.

📖 Plasma
Liquid component of the blood which contains plasma protein.

(a)

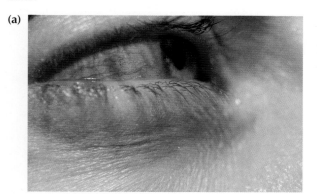

(b)

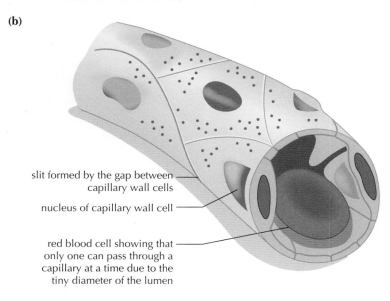

slit formed by the gap between capillary wall cells

nucleus of capillary wall cell

red blood cell showing that only one can pass through a capillary at a time due to the tiny diameter of the lumen

Figure 2.5.4 *(a) Tiny capillaries showing up in a person with an eye infection (b) The structure of a capillary – the wall is one cell thick and the lumen diameter is tiny*

Structure and function of veins

Veins have an outer layer of connective tissue containing elastic fibres but a much thinner muscular wall than arteries. They contain **valves** at intervals, as shown in **Figure 2.5.5**, which prevent the backflow of

PHYSIOLOGY AND HEALTH

Hint

Remember Veins have Valves.

blood which is a risk since the blood is at very low pressure. These valves come under extreme strain, especially in the legs where blood pressure is low and blood struggles to return to the heart against gravity. This can lead to the valve areas stretching through continual engorgement with blood, resulting in varicose veins as shown in **Figure 2.5.6**.

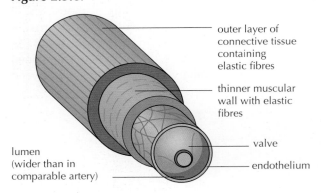

outer layer of connective tissue containing elastic fibres

thinner muscular wall with elastic fibres

valve

endothelium

lumen (wider than in comparable artery)

Figure 2.5.5 *Structure of a vein*

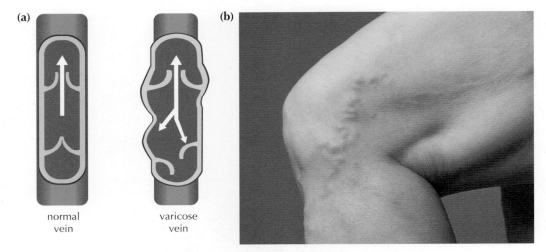

(a)

normal vein

varicose vein

(b)

Figure 2.5.6 *(a) The cause and (b) effect of varicose veins*

The table gives a comparison of the structure and functions of the different blood vessels.

Blood vessel	Function	Blood pressure	Wall structure	Diameter of lumen	Presence of valves
Artery	Carries blood away from the heart	High	Outer layer with elastic fibres and a middle layer with smooth muscle and more elastic fibres	Relatively narrow	No valves
Capillary	Form networks to allow exchange of material between blood and tissues	Pressure decreases as blood loses fluid to tissues	One cell thick with slits between loosely packed cells	Microscopically narrow – just wide enough for a red blood cell to pass through	No valves
Vein	Carries blood towards the heart	Low	Outer layer containing elastic fibres but middle layer with less smooth muscle than arteries	Relatively wide	Valves present

Control of blood flow

The muscular walls of arteries can contract and relax, producing changes in the diameter of the vessels. Vasoconstriction is reduction of the vessel diameters due to contraction of the smooth muscles in the wall. Vasodilation is the increase of the vessel diameters as the smooth muscles in the walls relax. Changes in vessel diameter change the volume of blood passing through the vessel. This can help in the control of body temperature as shown in **Figure 2.5.7**. When small arteries in skin dilate, more warm blood flows to the skin surface. The skin flushes and heat can be lost to the environment. If these arteries start to constrict, less blood flows to the skin surface. The skin pales and less heat is lost.

(a) Small arteries near to skin surface dilate and more warm blood flows near to the skin surface

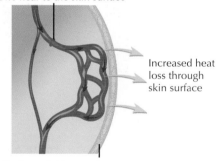

Increased heat loss through skin surface

Surface of skin may become flushed with extra blood

(b) Small arteries near to skin surface constrict and less warm blood flows near to the skin surface

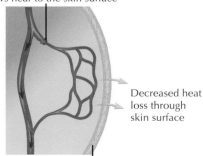

Decreased heat loss through skin surface

Surface of skin may become pale with less blood

Figure 2.5.7 *Thermal effects of (a) vasodilation and (b) vasoconstriction on blood flow to skin*

📖 **Tissue fluid**

Fluid which bathes cells in tissues; derived from blood plasma but lacking in plasma proteins.

📖 **Lymph**

Fluid made up from tissue fluid collected into lymph vessels, which circulates the body driven by the action of skeletal muscles.

Plasma, tissue fluid and lymph

Blood entering capillary networks from arteries is at fairly high pressure. This pressure can squeeze liquid out of the blood through the fine slits between capillary wall cells. This action is called **pressure filtration**. The liquid which is squeezed out is similar to plasma but lacks the plasma protein molecules which are too large to fit through the slits. The fluid is called **tissue fluid** and bathes the tissue cells around capillary networks. Tissue fluid contains dissolved glucose, oxygen and other substances which are supplied to the tissue cells by diffusion. Carbon dioxide and other metabolic waste diffuses out of the cells and into the tissue fluid to be excreted. Much of the tissue fluid, now low in nutrients and oxygen but high in metabolic waste and carbon dioxide, returns to the bloodstream. **Lymphatic vessels** absorb excess tissue fluid and return it as **lymph** to the circulatory system as shown in **Figure 2.5.8**.

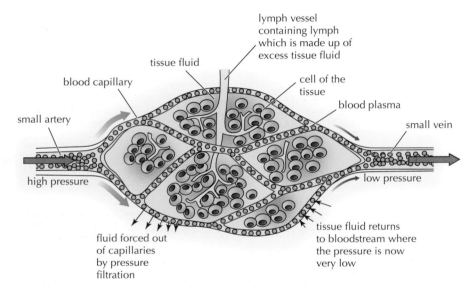

Figure 2.5.8 *The relationship between plasma, tissue fluid and lymph in a capillary network*

✦ **Make the link**

There is more about lymphocytes in Chapter 3.6 on page 231.

The lymph circulates around the body in lymph vessels squeezed along by the muscular movement of the body. It eventually returns to the bloodstream close to where blood returns to the right atrium as shown in **Figure 2.5.9**. Returning lymph passes through a series of lymph nodes where the lymphocytes and other immune system cells help to filter and clean the fluid.

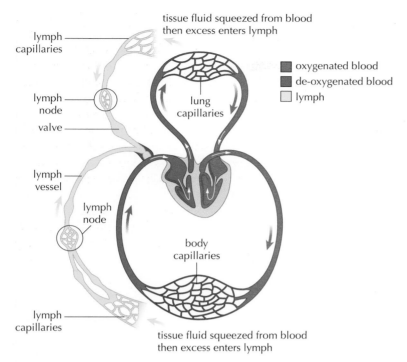

Figure 2.5.9 *Relationship between the circulation of lymph and blood. Note that the lymph circulates due to the action of skeletal muscles and not the heart*

GO! Activity 2.5.1 Work individually to . . .

Structured questions

1. Blood is pumped from the heart and circulates in arteries, capillaries and veins before returning to the heart.

    Describe the following vessels in terms of the relative diameters of their walls and lumens:

    a) Artery   2

    b) Capillary   2

    c) Vein   2

2. **Name** the tissues which make up the walls of arteries and veins.   2

3. **Describe** the blood pressure in arteries compared with veins.   1

4. **Describe** pressure filtration in capillaries.   2

5. **Give one** difference between the composition of plasma and tissue fluid.   1

6. **Describe** how excess tissue fluid is removed from capillary beds.   1

(continued)

7. Graphs a–d show information about various aspects of the circulation of blood through the vessels of the circulatory system.

(a)

(b)

(c)

(d)

a) **Identify** the vessels with the largest diameter. — 1

b) **Identify** the vessels with the highest total cross-sectional area. — 1

c) Use values from graph (c) to **describe** the changes in blood pressure as blood passes through the various parts of the circulatory system. — 3

d) **Explain** how vasodilation of small arteries allows blood flow to organs to be increased. — 3

e) **Explain** why the slow flow rate of blood through the capillaries is an advantage to their function. — 2

f) **Describe** the relationship between blood pressure and flow rate as blood passes through the veins and into the vena cava. — 1

Extended response questions

1. **Give an account** of the structure and function of arteries and veins. — 9

2. **Give an account** of the structure and function of capillaries. — 4

GO! Activity 2.5.2 Work in pairs to …

1. Practical activity: Looking at blood vessels

You will need: a microscope and a prepared slide of cross sections through blood vessels (these will not be human preparations).

Method

- Set up your microscope as directed by your teacher.
- Place the prepared slide onto the stage and focus on the slide at low power. If you can't do this, use the photograph below.

Make a labelled drawing of the two main vessels you can see. Make sure you include and label the following: artery, vein, relatively wide lumen, relatively narrow lumen, thick muscular wall, thin muscular wall, endothelium. Glue or tape your diagram into your notes.

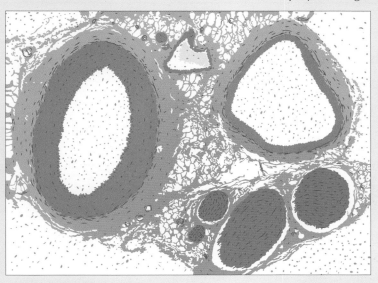

2. Flashcard activity

You will each need: a set of blank flashcards (A7 cards) and a stopwatch.

- Find the glossary terms for this chapter – they are the **black** typeface and red typeface terms. Using your blank cards, you should each make a set of flashcards for these terms – write the term on one side and the definition on the other. You will find the definitions in the chapter.
- Shuffle your cards and lay them out in a column, some showing terms and some showing definitions – you decide. Your partner should match their cards with yours, laying their cards in a column beside yours to give the corresponding term or definition. Time how long they take to do this.
- Now swap roles – your partner should lay out their cards and you should try the matching exercise while your partner times you.
- You should each keep your set of flashcards as a revision tool for later.

GO! Activity 2.5.3 Work as a group to ...

1. **Design and make a model of circulation and each of the blood vessel types.**

   You will need: 5 cm lengths of rubber vacuum hose, rubber Bunsen hose and drinking straw, a piece of A3 card, a glue stick and marker pens.

   Work together to design and make a labelled relief model to show the circulation of blood through the different blood vessels.

   Be prepared to describe your model to the class in terms of the names of the blood vessels and the order in which the blood passes through them, as well as a brief description of their walls and lumens.

Learning checklist

After working on this chapter, I can:

Knowledge and understanding

1. State that blood circulates from the heart through the arteries to the capillaries then to the veins and back to the heart. ◯ ◯ ◯

2. State that there is a decrease in blood pressure as blood moves away from the heart. ◯ ◯ ◯

3. State that blood vessels are tubes with walls composed of different tissues dependent on the function of the vessel. ◯ ◯ ◯

4. State that the central space or cavity of the blood vessels is called the lumen. ◯ ◯ ◯

5. State that the lumen is lined with a layer of cells called the endothelium. ◯ ◯ ◯

6. State that the endothelium lining the central lumen of blood vessels is surrounded by layers of tissue that differ between arteries, capillaries and veins. ◯ ◯ ◯

7. State that arteries carry blood away from the heart. ◯ ◯ ◯

8. State that blood is pumped through arteries at a high pressure. ◯ ◯ ◯

9. State that arteries have an outer layer of connective tissue containing elastic fibres and a thick middle layer containing smooth muscle with more elastic fibres. ◯ ◯ ◯

10. State that the thick elastic walls of the arteries stretch and recoil to accommodate the surge of blood after each contraction of the heart. ◯ ◯ ◯

11. State that the smooth muscle in the walls of small arteries can contract or relax, causing vasoconstriction and vasodilation to control blood flow. ◯ ◯ ◯

12. State that the ability of the small arteries to vasoconstrict or vasodilate allows the changing demands of the body's tissues to be met. ◯ ◯ ◯

13. State that during exercise, the small arteries supplying the muscles vasodilate, which increases the blood flow.

14. State that during exercise the small arteries supplying the abdominal organs vasoconstrict, which reduces the blood flow to them.

15. State that capillary walls are only one cell thick, which allows quick and efficient exchange of substances with tissues through their thin walls.

16. State that veins carry blood towards the heart.

17. State that veins have an outer layer of connective tissue containing elastic fibres but a much thinner muscular wall than arteries.

18. State that the central lumen of a vein is relatively wider than that of an artery.

19. State that valves are present in veins to prevent the backflow of blood.

20. State that valves are needed as the blood is flowing back to the heart at low pressure and generally against the force of gravity.

21. State that pressure filtration causes plasma to pass through capillary walls into the tissue fluid surrounding the cells.

22. State that tissue fluid and blood plasma are similar in composition, with the exception of plasma proteins, which are too large to be filtered through the capillary walls and are therefore not found in tissue fluid.

23. State that tissue fluid contains glucose, oxygen and dissolved substances which supply the tissues with all their requirements.

24. State that useful molecules such as glucose and oxygen diffuse into cells and carbon dioxide and other metabolic waste substances diffuse out of cells and into the tissue fluid to be excreted.

25. State that much of the tissue fluid re-enters the capillaries and returns to the blood.

26. State that lymphatic vessels absorb excess tissue fluid and return it as lymph to the circulatory system.

Skills

1. *Select information from line graphs.*

2. *Draw conclusions from line graphs.*

2.6 The structure and function of the heart

You should already know:

- The pathway of oxygenated and deoxygenated blood through heart, lungs and body.
- The structure of the heart, including the right and left atria, the right and left ventricles and the locations of the four valves.
- The functions of the atria, ventricles and valves.
- The location and functions of the associated blood vessels: the aorta, vena cava, pulmonary artery, pulmonary vein and coronary arteries.

Learning intentions

- Describe the flow of blood through the heart and its associated blood vessels.
- Describe cardiac output and how to calculate it.
- Describe the events of the cardiac cycle.
- Explain changes in blood pressure in the cardiac cycle and the action of valves.
- Describe the conducting system of the heart.

 • Describe the measurement of blood pressure and pulse rate.

Blood flow through the heart and its associated blood vessels

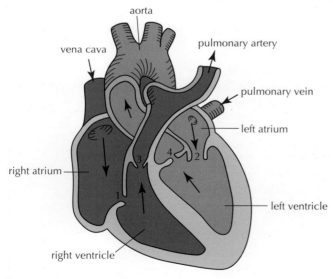

Figure 2.6.1 *The four chambers, four associated blood vessels and positions of the four valves of the heart:*
1 and 2 - atrio-ventricular (AV) valves
3 and 4 - semi-lunar (SL) valves

The heart is a hollow muscular organ made up of **cardiac muscle** tissue. It has four chambers, two **atria** (single: atrium) which receive blood from the veins and two **ventricles** which receive blood from the arteries. The right side of the heart receives deoxygenated blood (shown in blue) from the **vena cava** and delivers it out through the **pulmonary artery** to the lungs. The left side of the heart receives oxygenated blood (shown in red) from the lungs and delivers it out through the **aorta** to be circulated round the body as shown in **Figure 2.6.1**. The heart has four **valves** indicated by the numbers 1–4 in the diagram. These valves operate in a coordinated way to prevent the backflow of blood during circulation.

Cardiac output and its calculation

The volume of blood pumped through each ventricle per minute is the **cardiac output (CO)**. Cardiac output is determined by multiplying the heart rate (HR) by the stroke volume (SV).

CO (volume per minute) = HR (beats per minute) × SV (volume of blood expelled in a beat)

The left and right ventricles pump equal volumes of blood through the aorta and pulmonary artery respectively.

The cardiac cycle

The **cardiac cycle** is the events which occur in the heart during a single heartbeat. The cycle has phases of relaxation of the heart muscle which are called **diastole**. Phases in the cycle during which heart muscle is contracted are called **systole** and systole is divided into atrial systole and ventricular systole.

During diastole, blood which has returned to the atria flows on into the ventricles. This is followed by atrial systole which forces the remainder of the blood through the atrio-ventricular (AV) valves to the ventricles. During ventricular systole, increased blood pressure causes the AV valves to close and the blood is pumped out through the semi-lunar (SL) valves to the aorta and pulmonary artery. As diastole sets in again, the pressure remaining in the arteries closes the SL valves and blood starts to re-enter the atria. **Figure 2.6.2** shows the main phases of the cardiac cycle and the average times taken.

> 📖 **Cardiac output (CO)**
>
> Volume of blood expelled from one ventricle of the heart per minute.

> 📖 **Diastole**
>
> Phase of the cardiac cycle during which cardiac muscle is relaxed.

> 📖 **Systole**
>
> Phase of the cardiac cycle during which cardiac muscle is contracted.

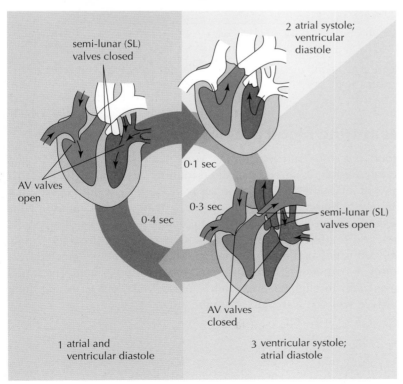

Figure 2.6.2 *Phases of the cardiac cycle, their features and average durations*

Blood pressure and heart sounds

Blood pressure changes during the cardiac cycle are responsible for the opening and closing of the heart valves. At the start of ventricular systole the AV valves are forced to close and at the end of ventricular systole the SL valves are forced to close. The closing of the AV and SL valves produces the heart sounds heard with a stethoscope.

The closing of the AV valves is a deeper, longer sound sometimes written as LUBB and the closing of the SL valves is a higher pitched and shorter sound sometime written as DUP. **Figure 2.6.3** shows the blood pressure changes in the aorta and the left ventricle during an average cardiac cycle. Note that ventricular systole starts and ends with a heart sound.

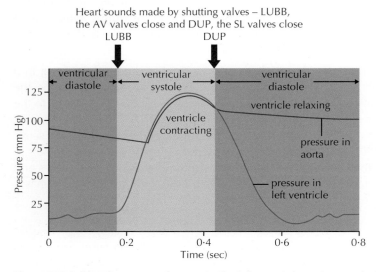

Figure 2.6.3 *Blood pressure changes in the left ventricle and aorta during a cardiac cycle*

Measuring blood pressure and pulse rate

Blood pressure increases during ventricular systole and decreases during diastole. It can be measured routinely using a device called a **sphygmomanometer**. Digital versions of these exist but **Figure 2.6.4** shows a traditional sphygmomanometer being used. A rubber cuff is inflated around the upper arm to a pressure greater than the systolic pressure in the upper arm. This stops blood flow in the artery there. The cuff is deflated gradually and blood starts to flow (detected by a pulse) at systolic pressure. This can be read off on a scale. The blood flows freely through the artery (and a pulse is not detected) at diastolic pressure which is also read off on the scale.

📖 **Sphygmomanometer**

Device for measuring blood pressure in millimetres of mercury (mm Hg).

(a) A cuff is placed round an artery in the upper arm and inflated to stop blood flow; the blood pressure shows on a scale and a sthethoscope detects no sounds

(b) The cuff is slowly deflated, blood starts to flow at systolic pressure which shows on the scale and a pulse in detected; the average systolic pressure for a young adult is 120mmHg

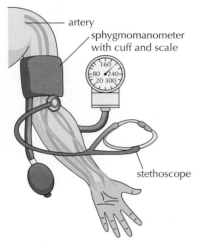

(c) As cuff is further deflated blood starts to flow freely and a pulse is not detected; the average diastolic pressure for a young adult is 80mmHg

Figure 2.6.4 *Using a traditional manual sphygmomanometer and a stethoscope to measure a patient's blood pressure*

A typical blood pressure reading for a young adult is 120/80 mm Hg. The first number is the higher systolic pressure and the lower number is the diastolic pressure. **Hypertension** is the term for high blood pressure and is a major risk factor for many diseases, including coronary heart disease (CHD).

Pulse rate can be measured externally by feeling or listening to arteries which pass over bones near to the surface of the skin at places called pressure points. It is mostly taken at the wrist where the radial artery passes over the end of bones of the lower arm. This can be done using the fingers to feel the wrist pulse and timing with a stopwatch or by using a **pulsometer**, as shown in **Figure 2.6.5**.

📖 Pulsometer

Device for measuring pulse rate in beats per minute (bpm).

(a)

(b)

Figure 2.6.5 *(a) Taking a patient's pulse rate manually (b) Using a sports pulsometer linked to a mobile phone to measure pulse rate*

📖 Auto-rhythmic cells

Cells within the sino-atrial node (SAN) which have an intrinsic rhythm and set the rate of heart contractions.

📖 Sino-atrial node (SAN)

Cluster of auto-rhythmic cells in the wall of the right atrium which dictate the rate of heart contractions; receives nerve impulse from the medulla.

📖 Atrio-ventricular node (AVN)

Cluster of nerve cells found at the junction of the atria and ventricles of the heart; passes impulses from the SAN to the ventricle walls.

🔎 Hint

Remember that the conducting system is about nerves running through the heart **NOT** blood running through it.

📖 Sympathetic nerve

Nerve fibre that stimulates an acceleration of the heart rate; releases noradrenaline.

📖 Parasympathetic nerve

Nerve fibre that stimulates a decrease in heart rate; releases acetylcholine.

The structure and function of the cardiac conducting system of the heart

The heartbeat originates in the heart itself. The **auto-rhythmic cells** of the **sino-atrial node (SAN)** or pacemaker, located in the wall of the right atrium, set the rate at which the heart contracts.

The timing of cardiac muscle cell contraction is controlled by impulses from the SAN spreading through the atria causing atrial systole. They then travel to the **atrio-ventricular node (AVN)**, located in the centre of the heart. Impulses from the AVN travel down a bundle of fibres in the central wall of the heart which then divides to transfer impulses up through the walls of both ventricles, causing ventricular systole, as shown in **Figure 2.6.6**.

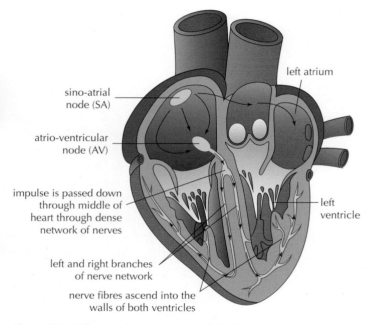

left atrium

sino-atrial node (SA)

atrio-ventricular node (AV)

impulse is passed down through middle of heart through dense network of nerves

left ventricle

left and right branches of nerve network

nerve fibres ascend into the walls of both ventricles

Figure 2.6.6 *The conducting system of the heart*

Control of heart rate

The heart rate is regulated by the **medulla** in the brain, which sends nervous impulses to the sino-atrial node (SAN) through the antagonistic action of the autonomic nervous system (ANS).

A **sympathetic nerve** releases **noradrenaline**, which increases the heart rate, whereas a **parasympathetic nerve** releases **acetylcholine**, which decreases the heart rate, as shown in **Figure 2.6.7**. It is important that the heart rate varies to ensure that the supply of oxygenated blood to tissues keeps pace with the demands of the changing activity of the body.

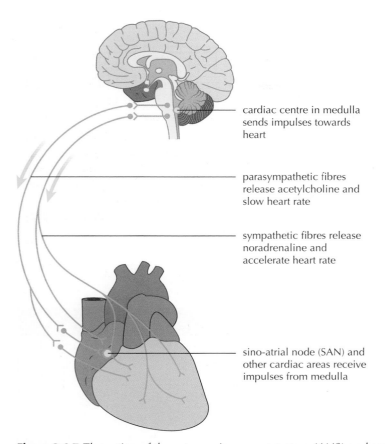

Make the link

There is more about the medulla in Chapter 3.2 on page 202.

cardiac centre in medulla sends impulses towards heart

parasympathetic fibres release acetylcholine and slow heart rate

sympathetic fibres release noradrenaline and accelerate heart rate

Make the link

There is more about sympathetic and parasympathetic nerve fibres in Chapter 3.1 on page 195.

sino-atrial node (SAN) and other cardiac areas receive impulses from medulla

Figure 2.6.7 *The action of the autonomic nervous system (ANS) on heart rate*

Electrocardiograms (ECGs)

Impulses in the heart generate electrical currents that can be detected by an electrocardiogram machine. The results are displayed on an **electrocardiogram (ECG)** as shown in **Figure 2.6.8** and produced so that heart rate and details of the waves produced at atrial systole, ventricular systole and diastole can be seen and studied. The letters P, Q, R, S and T are used to indicate points of reference on the trace. P indicates activity connected to atrial systole; Q, R and S indicate ventricular systole and T is the heart's recovery before the next beat. Abnormal ECGs can indicate problems with the heart and are useful in diagnosis, as shown in **Figure 2.6.9**.

Electrocardiogram (ECG)

A visual trace of the electrical activity of the heart which can be used to detect abnormalities.

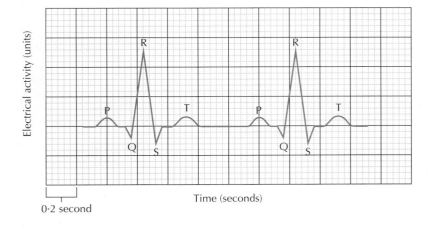

Figure 2.6.8 *Normal electrocardiogram (ECG) showing a trace for two heartbeats – note the timescale from which the heart rate could be calculated*

Make the link

There is more about health issues related to the heart and circulation in Chapter 2.7 on page 165.

ventricles contracting more rapldly than normal
fast heartbeat

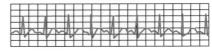

ventricles contracting slower than normal
slow heartbeat

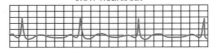

ventricles contracting irregularly
irregular heartbeat

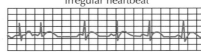

Figure 2.6.9 *Normal electrocardiogram (ECG) with some common abnormalities*

GO! Activity 2.6.1 Work individually to ...

Structured questions

1. **Give** the functions of the following heart valves:
   a) AV valves   1
   b) SL valves   1
2. The heart has a system of nodes joined by nerves which run through its structure.
   Give the location and function of the:
   a) sino-atrial node (SAN)   2
   b) AV node   2
3. The cardiac cycle has phases called systole and diastole.
   Describe the state of cardiac muscle during each of these phases.   2
4. **Copy and complete** the following table showing data on cardiac output for some different individuals. Use the formula for cardiac output given on page 153.

Individual	Pulse rate (bpm)	Stroke volume (cm³)	Cardiac output (cm³ per minute)
1	60		4,500
2		80	4,000
3	70	85	
4	75		5,700
5		80	6,400
6	85	76	
7	72		4,968
8	55	71	

5. A clinical trial was carried out to test the effects of a new drug on blood pressure.

The resting blood pressure of eight participants was measured. The participants were each given a dosage of the drug by an oral tablet and one hour later their blood pressure was remeasured. The results are shown in the table.

Participant	Initial blood pressure (mm Hg)		Blood pressure one hour after taking the oral tablet (mm Hg)	
	Systolic	Diastolic	Systolic	Diastolic
1	125	85	133	93
2	122	82	129	92
3	130	90	127	89
4	115	75	131	88
5	110	70	135	94
6	123	83	132	91
7	117	77	125	86
8		78	128	87
Average reading	120	80	130	90

a) i. **Identify** the independent variable in this trial.   1

   ii. **Identify one** variable which should be the same during this trial.   1

b) i. **Calculate** the percentage increase in average diastolic pressure one hour after taking the tablet.   1

   ii. **Calculate** the missing systolic pressure value for Participant 8 before taking the drug.   1

c) On a separate piece of graph paper, **draw a bar chart** to show the average blood pressure readings shown in the table.   2

d) **Describe** a control which should have been included in this trial.   1

e) It was concluded that the drug caused increases in blood pressure.

   Explain why this conclusion is not reliable.   1

f) **Name** an instrument which can be used to measure blood pressure.   1

☀ Make the link

There is more about clinical trials in Chapter 3.8 on page 247. There is more about variables, presenting information and concluding in Chapter 4 on page 263.

(continued)

6. The graph below shows the stroke volume and heart rate for an individual exercising on a treadmill as the gradient of the treadmill was gradually raised from 0 to 30 degrees.

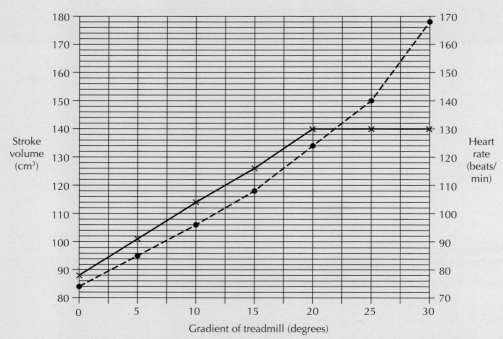

a) **Use values from the graph to describe** the changes in the individual's stroke volume as the gradient of the treadmill was raised from 0 to 30 degrees. 2

b) **Calculate** the increase in heart rate as the gradient of the treadmill was raised from 5 degrees to 15 degrees. 1

c) **Identify** the stroke volume of the individual when their heart rate was 124 beats per minute. 1

d) **Calculate** the cardiac output of this individual when the treadmill was set at 25 degrees. 1

e) **Calculate** the average increase in stroke volume per degree of increase in gradient of the treadmill. 1

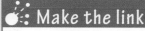 Make the link

There is more about selecting and processing information from a double axis line graph in Chapter 4 on page 267.

Extended response questions

1. **Give an account** of the conducting system of the heart. 5
2. **Give an account** of blood flow through the heart, naming the chambers, valves and blood vessels through which it passes. 8

GO! Activity 2.6.2 Work in pairs to ...

1. **Card sequencing activity: Cardiac cycle**

   You will need: a stopwatch and a mini whiteboard with a marker pen.

   - Read the phrases in the grid below which describe events in the cardiac cycle.

   - Work together to arrange the phrases in the correct order – start with phrase 2. Mark your answer on the whiteboard. Your teacher will check your work or put the correct sequence on the board.

   - You should each try this again individually against the clock – who is faster to the correct answer?

   - Your teacher may provide you with a photocopy of the grid. If so, cut it into strips, re-sequence it and glue it into your notes to make a permanent flowchart.

1 Blood flows into the ventricles	**4** During ventricular systole, blood is forced through the SL valves
2 During atrial systole, blood is forced through the AV valves	**5** The SL valves close
3 At diastole, the heart muscle relaxes	**6** The AV valves close and blood is forced into the main arteries

2. **Research electrocardiograms (ECGs).**

   Visit the NHS web page link below:

   www.nhs.uk/conditions/electrocardiogram/

   Find out what is meant by the following conditions which can be indicated by abnormal ECG readings:

   - Arrhythmia
   - Coronary heart disease
   - Heart attack
   - Cardiomyopathy

3. **Flashcard activity**

   You will each need: a set of blank flashcards (A7 cards) and a stopwatch.

   - Find the glossary terms for this chapter – they are the **black** typeface and **red** typeface terms. Using your blank cards, you should each make a set of flashcards for these terms – write the term on one side and the definition on the other. You will find the definitions in the chapter.

   - Shuffle your cards and lay them out in a column, some showing terms and some showing definitions – you decide. Your partner should match their cards with yours, laying their cards in a column beside yours to give the corresponding term or definition. Time how long they take to do this.

   - Now swap roles – your partner should lay out their cards and you should try the matching exercise while your partner times you.

   - You should each keep your set of flashcards as a revision tool for later.

GO! Activity 2.6.3 Work as a group to ...

1. **Class practical activity: using a pulsometer**

   You will need: pulsometers or stopwatches.

   ⚠ Your teacher will give specific safety instructions for carrying out this experiment and will give directions for the use of pulsometers or the taking of pulse rates manually, if required.

   ### Method

   - Each member of the class should find their resting pulse rate using a pulsometer or by having a partner take it manually.
   - Add your own value to results on the board.
   - Calculate the average reading for the class.
   - Your teacher may ask you to repeat the measurements after taking some mild exercise such as jogging a set distance in the school grounds.
   - Add your new results to the board and calculate the average reading for the class after taking the exercise.
   - You should each write a short (100 words) report saying what you did and the average results you obtained. Include a conclusion and a prediction of change in pulse rate if the activity were repeated with more vigorous exercise.

Learning checklist

After working on this chapter, I can:

Knowledge and understanding

1. State that the heart has four chambers (right atrium, right ventricle, left atrium and left ventricle) and works as a double pump. ⬭ ⬭ ⬭

2. State that the right side collects deoxygenated blood from the body and pumps it to the lungs to collect oxygen. ⬭ ⬭ ⬭

3. State that the left side collects oxygenated blood from the lungs and pumps it to the body. ⬭ ⬭ ⬭

4. State that the volume of blood pumped through each ventricle per minute is the cardiac output (CO). Cardiac output is determined by heart rate (HR) and stroke volume (SV) (CO = HR × SV). ⬭ ⬭ ⬭

5. State that the left and right ventricles pump the same volume of blood through the aorta and pulmonary artery. ⬭ ⬭ ⬭

6. State that the cardiac cycle is the pattern of contraction (systole) and relaxation (diastole) of the heart muscle in one complete heartbeat. ⬭ ⬭ ⬭

7. State that during diastole, blood returning to the atria flows into the ventricles. ⬭ ⬭ ⬭

8. State that atrial systole transfers the remainder of the blood through the atrio-ventricular (AV) valves to the ventricles.

9. State that ventricular systole closes the AV valves and pumps the blood out through the semi-lunar (SL) valves to the aorta and pulmonary artery.

10. State that in diastole, the higher pressure in the arteries closes the SL valves.

11. State that deoxygenated blood returning from the body via the vena cava fills the right atrium during atrial diastole.

12. State that the build-up of pressure during atrial diastole forces open the atrio-ventricular (AV) valves and blood flows into the right ventricle during ventricular diastole. The right atrium contracts, forcing all the blood into the right ventricle (atrial systole).

13. State that once full, the right ventricle's muscular walls contract (ventricular systole), closing the AV valves and forcing the blood up through the semi-lunar valves and out through the pulmonary artery to the lungs.

14. State that oxygenated blood returning from the lungs via the pulmonary vein fills the left atrium during atrial diastole.

15. State that the build-up of pressure during atrial diastole forces open the atrio-ventricular (AV) valves and blood flows into the left ventricle. The left atrium contracts, forcing all the blood into the left ventricle (atrial systole).

16. State that once full, the left ventricle's muscular walls contract (ventricular systole), closing the AV valves and forcing the blood up through the semi-lunar (SL) valves and out through the aorta to the body's organs.

17. State that the opening and closing of the AV and SL valves produces the heart sounds heard with a stethoscope.

18. State that the heartbeat originates in the heart itself.

19. State that the auto-rhythmic cells of the sino-atrial node (SAN) or pacemaker, located in the wall of the right atrium, set the rate at which the heart contracts.

20. State that the timing of cardiac muscle cell contraction is controlled by impulses from the SAN spreading through the atria causing atrial systole.

21. State that the impulses then travel to the atrio-ventricular node (AVN), located in the centre of the heart. Impulses from the AVN travel down fibres in the central wall of the heart and then up through the walls of the ventricles, causing ventricular systole.

22. State that impulses in the heart generate currents that can be detected by an electrocardiograph (ECG).

23. State that the medulla in the brain regulates the rate of the sino-atrial node (SAN) through the antagonistic action of the autonomic nervous system (ANS).

24. State that a sympathetic nerve releases noradrenaline, which increases the heart rate, whereas a parasympathetic nerve releases acetylcholine, which decreases the heart rate.

25. State that blood pressure increases during ventricular systole and decreases during diastole.

26. State that measurement of blood pressure is performed using a sphygmomanometer.

27. State that an inflatable cuff stops blood flow in the artery and deflates gradually.

28. State that the blood starts to flow (detected by a pulse) at systolic pressure.

29. State that the blood flows freely through the artery (and a pulse is not detected) at diastolic pressure.

30. State that a typical blood pressure reading for a young adult is 120/80 mm Hg.

31. State that hypertension (high blood pressure) is a major risk factor for many diseases and conditions, including coronary heart disease.

Skills

1. *Plan procedures by identifying variables and controls.*

2. *Evaluate the reliability of procedures.*

3. *Present information as a bar chart.*

4. *Process information by calculating percentages.*

5. *Process information by calculating cardiac output.*

2.7 Pathology of cardiovascular disease (CVD)

You should already know:

- The location and function of the coronary arteries.

Learning intentions

- Describe the process of atherosclerosis.
- Describe the process of thrombosis.
- Describe the causes and effects of peripheral vascular disorders.
- Describe the control of cholesterol in the body.

The process of atherosclerosis

Normal arteries have smooth endothelium with no obstructions to normal blood flow. In **atherosclerosis**, fatty substances consisting mainly of **cholesterol**, fibrous material and calcium accumulate and bind together to form **atheromas** or plaques beneath the endothelium of the artery as shown in **Figure 2.7.1**.

As the atheroma grows the artery wall thickens and loses its elasticity. The diameter of its lumen becomes narrowed and blood flow becomes restricted, which increases blood pressure.

Atherosclerosis is the root cause of various **cardiovascular diseases (CVDs)**, including angina, heart attacks, strokes and peripheral vascular disease.

📖 Atherosclerosis

Development of atheromas in arteries leading to hardening and blockage of the arteries; root cause of various cardiovascular diseases.

📖 Cholesterol

Lipid needed for cell membranes and in the synthesis of steroid hormones; can contribute to atherosclerosis if present in excess in bloodstream.

📖 Cardiovascular disease (CVD)

Various diseases affecting the functioning of the heart and circulation.

(a)

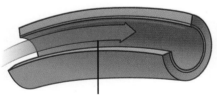

blood flows smoothly through

cross section shows unobstructed lumen

(b)

atheroma

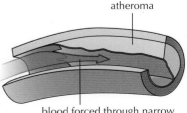

blood forced through narrow channel increasing blood pressure

cross section shows lumen narrowed by atheroma

Figure 2.7.1 (a) Normal artery (b) Artery with an atheroma – note how the atheroma has narrowed the artery lumen, forcing blood through a narrower channel and raising the blood pressure

165

Thrombosis and embolism

If an atheroma ruptures and breaks from an artery wall, the endothelium may be damaged. Blood **clotting factors** may be released which can trigger a cascade of chemical reactions in the area. A substance called **prothrombin** can be converted to **thrombin** which in turn converts the soluble plasma protein **fibrinogen** to **fibrin**. Fibrin is insoluble and forms fibres which trap red blood cells into a clot called a **thrombus** as shown in **Figure 2.7.2**. A thrombus may seal the endothelium wall in the damaged area and provide a scaffold for scar tissue, resulting in a blockage to the blood vessel as shown in **Figure 2.7.3**.

clotting factors stimulate conversion of prothrombin to thrombin

thrombin causes conversion of soluble fibrinogen to insoluble, fibrous fibrin

red blood cells become trapped in network of fibres

fibres of insoluble fibrin form a network encouraging a thrombus to develop

Figure 2.7.2 *Formation of a thrombus as a result of the release of clotting factors at a site of endothelial damage*

Thrombosis in a coronary artery may lead to a **myocardial infarction (MI)**, commonly known as a heart attack, as shown in **Figure 2.7.3**. The blockage causes oxygen deprivation to areas of cardiac muscle which then stop working. Thrombosis in an artery in the brain may lead to a stroke.

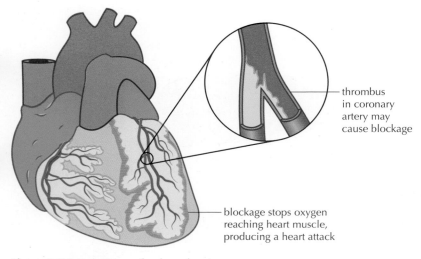

thrombus in coronary artery may cause blockage

blockage stops oxygen reaching heart muscle, producing a heart attack

Figure 2.7.3 *Formation of a thrombus in a coronary artery may cause complete blockage and a heart attack*

blood vessel narrowed by atheroma but blood flows through to tissues beyond

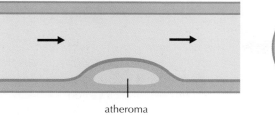
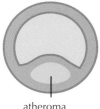

atheroma   atheroma

atheroma ruptures and a thrombus forms, blocking the vessel and preventing blood flow to tissues beyond

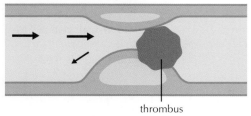
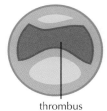

thrombus   thrombus

Figure 2.7.4 *Development of a thrombus following damage caused by ruptured atheroma*

If a thrombus or part of a thrombus breaks loose as shown in **Figure 2.7.5**, it forms an **embolus** which may travel from the area through blood vessels and may eventually block one of them. This could stop oxygenated blood reaching an area of the body, resulting in oxygen deprivation and the death of tissues.

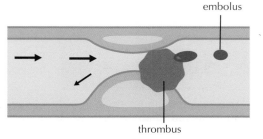

embolus

thrombus

Figure 2.7.5 *Formation of an embolus which has broken loose and is free to travel in the bloodstream and may eventually block a smaller blood vessel*

Causes and effects of peripheral vascular disorders

Peripheral vascular disease is caused by the narrowing of the arteries, other than those of the heart or brain, due to atherosclerosis. The arteries to the legs are most commonly affected. Pain is experienced in the leg muscles due to a limited supply of oxygen. A **deep vein thrombosis (DVT)** is a blood clot that forms in a deep vein, most commonly in the leg. This can break off as shown in **Figure 2.7.6** and result in a pulmonary embolism in the lungs.

📖 **Peripheral vascular disease**

Condition caused by blockage to arteries other than coronary arteries, the aorta or those in the brain.

📖 **Deep vein thrombosis (DVT)**

A blood clot in a vein deep into the body, very often in the leg.

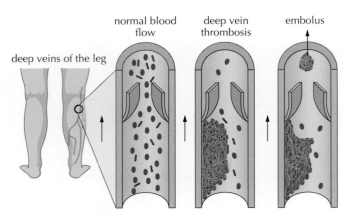

Figure 2.7.6 *Deep vein thrombosis (DVT) – a thrombus in a leg vein reduces the flow of blood, causing oxygen deprivation to muscles and pain. If an embolus breaks off it can lodge in pulmonary blood vessels in the lungs*

Control of cholesterol levels in the body

Cholesterol

Cholesterol is a type of lipid found in the cell membrane. It is also used to make the sex hormones testosterone, oestrogen and progesterone.

Cholesterol is synthesised by all cells, although 25% of total production takes place in the liver. A diet high in saturated fats or cholesterol causes an increase in cholesterol levels in the blood.

The roles of high-density lipoproteins (HDL) and low-density lipoproteins (LDL)

High-density lipoproteins (HDL) in the bloodstream transport excess cholesterol from the body cells to the liver for elimination. Elimination prevents accumulation of cholesterol in the blood.

Low-density lipoproteins (LDL) transport cholesterol to body cells. Most cells have **LDL receptors** that take LDL into the cell where it releases cholesterol. Once a cell has sufficient cholesterol, a negative feedback system inhibits the synthesis of new LDL receptors and LDL circulates in the blood where it may deposit cholesterol in the arteries, forming atheromas.

Make the link

There is more about sex hormones in Chapters 2.1, 2.2, and 2.3 on pages 100, 107 and 116.

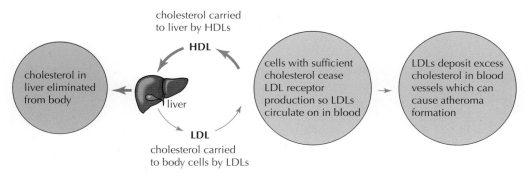

Figure 2.7.7 *Function of HDLs and LDLs – note that the presence of high levels of HDLs assists the removal of cholesterol and that high levels of LDLs can cause atheroma formation*

HDL : LDL

A higher ratio of HDL to LDL is beneficial to health because it results in lower blood cholesterol and a reduced chance of atherosclerosis. Higher ratios are achieved by eating lower levels of total fat and increasing the proportion of unsaturated as opposed to saturated fat in the diet. Physical activity tends to raise HDL level which in turn increases the ratio of HDL to LDL. A healthy lifestyle with a low fat diet and plenty of exercise is important.

Prescribed medicinal drugs such as **statins**, shown in **Figure 2.7.8**, reduce blood cholesterol by inhibiting the synthesis of cholesterol by liver cells.

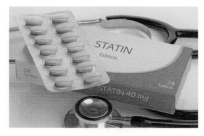

Figure 2.7.8 *Statins may be prescribed by a doctor – they inhibit the synthesis of cholesterol by liver cells and so can contribute to lowering blood cholesterol levels*

GO! Activity 2.7.1 Work individually to …

Structured questions

1. **Describe** how a thrombus forms.   3
2. **Describe** how an embolus forms.   2
3. **Name** the events which are caused by thrombosis in the following areas:
   a) Coronary artery   1
   b) Blood vessel in the brain   1
4. **Describe** the normal functions of cholesterol.   2
5. **Explain** why exercise and a low fat diet are good for the health of the circulatory system.   2
6. **Name** a drug which might be prescribed for high blood cholesterol levels.   1

(continued)

7. The table shows the death rates from cardiovascular disease (CVD) in populations of two Scottish cities between 2004 and 2010.

Year	Death rates from CVD (numbers per 100,000)			
	Population of City A		Population of City B	
	Males	Females	Males	Females
2004	630	450	610	410
2006	550	410	520	390
2008	430	340	410	350
2010	420	280	320	210

a) **Identify two** trends which can be seen in the data.　　　　　　　　　　2

b) **Calculate** the simple whole number ratio of deaths from CVD in the males and females in City A in 2010.　　　　　　　　　　1

c) **Calculate** the average decrease in deaths from CVD per year for males in City A.　　1

d) **Predict** the death rate for females in City B in 2012.　　　　　　　　　　1

e) **Explain** how the data allows valid comparisons between the two cities even though their populations are likely to be of different size.　　　　　　　　　　1

> ### ⚛ Make the link
>
> There is more about selecting and processing information and predicting from tables in Chapter 4 on page 267.

Extended response questions

1. **Give an account** of the roles of high-density lipoproteins (HDL) and low-density lipoproteins (LDL) in the regulation of blood cholesterol levels.　　　　　　4

2. **Write notes** on the causes of atherosclerosis and its effects on health.　　　8

ⒼⓄ Activity 2.7.2 Work in pairs to …

1. **Card sequencing activity: Events in thrombosis**

   You will need: a stopwatch and a mini whiteboard with a marker pen.

   • Read the phrases in the grid, which describe events in thrombosis.

   • Work together to arrange the phrases in the correct order – start with phrase 3. Mark your answer on the whiteboard. Your teacher will check your work or put the correct sequence on the board.

   • You should each try this again individually against the clock – who is faster to the correct answer?

- Your teacher may provide you with a photocopy of the grid. If so, cut it into strips, re-sequence it and glue it into your notes to make a permanent flowchart.

1 Prothrombin is converted to thrombin	4 Thrombin causes conversion of fibrinogen to fibrin
2 Fibrin fibres form a scaffold and trap red blood cells to produce a clot called a thrombus	5 Clotting factors are released
3 An atheroma ruptures	6 A thrombus may block an important artery

 2. Flashcard activity

You will each need: a set of blank flashcards (A7 cards) and a stopwatch.

- Find the glossary terms for this chapter – they are the **black** typeface and **red** typeface terms. Using your blank cards, you should each make a set of flashcards for these terms – write the term on one side and the definition on the other. You will find the definitions in the chapter.

- Shuffle your cards and lay them out in a column, some showing terms and some showing definitions – you decide. Your partner should match their cards with yours, laying their cards in a column beside yours to give the corresponding term or definition. Time how long they take to do this.

- Now swap roles – your partner should lay out their cards and you should try the matching exercise while your partner times you.

- You should each keep your set of flashcards as a revision tool for later.

GO! Activity 2.7.3 Work as a group to . . .

1. **Design and make a poster on atherosclerosis, thrombosis and embolism.**

   You will need: A3 paper or card and coloured pencils, or a mini whiteboard and marker pens.

   Make a poster to illustrate and explain the differences between atherosclerosis, thrombosis and embolism.

   Include accurate labelled diagrams of each condition and explain each condition in simple text.

   Include **two** pieces of advice for lowering the risk of these conditions.

 2. **Research cardiovascular disease (CVD).**

   Visit the NHS web page below:

   www.nhs.uk/conditions/cardiovascular-disease/

   Prepare a three- or four-slide PowerPoint presentation (you could prepare one slide each) covering the following:

   - the forms that CVD can take
   - the risk factors in CVD
   - the steps which people can take to reduce their risk of CVD

   Be prepared to present your slides to the class.

Learning checklist

After working on this chapter, I can:

Knowledge and understanding

1. State that atherosclerosis is the accumulation of fatty material (consisting mainly of cholesterol), fibrous material and calcium, forming an atheroma or plaque beneath the endothelium.

2. State that an atheroma forms beneath the endothelium (inner lining) of the artery wall.

3. State that as an atheroma grows, the artery thickens and loses its elasticity.

4. State that an atheroma reduces the diameter of the lumen of an artery, which restricts blood flow and results in increased blood pressure.

5. State that atherosclerosis is the root cause of various cardiovascular diseases (CVDs), including angina, heart attack, stroke and peripheral vascular disease.

6. State that atheromas may rupture, damaging the endothelium.

7. State that the damage to the endothelium releases clotting factors that activate a cascade of reactions resulting in the conversion of the enzyme prothrombin to its active form thrombin.

8. State that thrombin causes molecules of the plasma protein fibrinogen to form threads of fibrin.

9. State that the fibrin threads form a meshwork that clots the blood, seals the wound and provides a scaffold for the formation of scar tissue.

10. State that the formation of a clot (thrombus) is referred to as thrombosis.

11. State that in some cases a thrombus may break loose, forming an embolus which travels through the bloodstream until it blocks a blood vessel.

12. State that thrombosis in a coronary artery may lead to a myocardial infarction (MI), commonly known as a heart attack.

13. State that thrombosis in an artery in the brain may lead to a stroke.

14. State that thrombosis normally results in the death of some of the tissue served by the blocked artery as the cells are deprived of oxygen.

15. State that peripheral vascular disease is narrowing of the arteries due to atherosclerosis of arteries other than those of the heart or brain.

16. State that the arteries to the legs are most commonly affected by peripheral vascular disease.

17. State that with a deep vein thrombosis (DVT), pain is experienced in the leg muscles due to a limited supply of oxygen.

18. State that a deep vein thrombosis (DVT) is a blood clot that forms in a deep vein, most commonly in the leg.

19. State that a pulmonary embolism is caused by part of a thrombus breaking free and travelling through the bloodstream to the pulmonary artery, where it can cause a blockage, resulting in chest pain and breathing difficulties.

20. State that cholesterol is a type of lipid found in the cell membrane.

21. State that cholesterol is also used to make the sex hormones testosterone, oestrogen and progesterone.

22. State that cholesterol is synthesised by all cells, although 25% of total production takes place in the liver.

23. State that a diet high in saturated fats or cholesterol causes an increase in cholesterol levels in the blood.

24. State that lipoproteins contain lipid and protein.

25. State that high-density lipoprotein (HDL) transports excess cholesterol from the body cells to the liver for elimination. This prevents accumulation of cholesterol in the blood.

26. State that low-density lipoprotein (LDL) transports cholesterol to body cells.

27. State that most cells have LDL receptors that take LDL into the cell where it releases cholesterol.

28. State that once a cell has sufficient cholesterol a negative feedback system inhibits the synthesis of new LDL receptors and LDL circulates in the blood where it may deposit cholesterol in the arteries, forming atheromas.

29. State that a higher ratio of HDL to LDL will result in lower blood cholesterol and a reduced chance of atherosclerosis.

30. State that regular physical activity tends to raise HDL levels.

31. State that dietary changes aim to reduce the levels of total fat in the diet and to replace saturated with unsaturated fats.

32. State that drugs such as statins reduce blood cholesterol by inhibiting the synthesis of cholesterol by liver cells.

Skills

1. *Select information from a table.*

2. *Process information by calculating percentage change, average change and ratio from a table.*

3. *Predict from information in a table.*

2.8 Blood glucose levels and obesity

You should already know:

- Endocrine glands release hormones into the bloodstream.
- Hormones are chemical messengers.
- A target tissue has cells with complementary receptor proteins for specific hormones, so only that tissue will be affected by these hormones.
- Blood glucose regulation involves the hormones insulin and glucagon released by the pancreas.
- Insulin causes liver cells to store excess glucose as glycogen.
- Release of glucose from glycogen in the liver is stimulated by glucagon.

Learning intentions

- Describe the effects on the body of chronic elevated glucose levels.
- Explain the regulation of glucose by negative feedback.
- Describe the causes and effects of type 1 and type 2 diabetes.
- Describe obesity and its effects on health.
- Calculate BMI values from body mass and height measurements.

Blood glucose levels

Normal blood glucose levels range from less than 100 mg of glucose per dL of blood after several hours without food to under 140 mg of glucose per dL of blood two hours after a meal. Chronic elevation of blood glucose levels leads to endothelium cells lining the blood vessels taking in more glucose than normal which damages the cells and causes disruption to the endothelium layer. Atherosclerosis may develop at the site of damage leading to cardiovascular disease (CVD), stroke or peripheral vascular disease, as outlined in Chapter 2.7 on page 165. Small blood vessels damaged by elevated glucose levels may result in haemorrhage of blood vessels in the retina, renal failure or peripheral nerve dysfunction.

Regulation of blood glucose levels

Regulation of blood glucose depends on hormones produced by the **pancreas** and chemical reactions in the **liver**. The positions of the pancreas and liver are shown in **Figure 2.8.1**.

Following a meal, blood glucose levels rise. Receptors in the pancreas respond to raised blood glucose levels by increasing secretion of the hormone **insulin** from pancreatic cells. Insulin targets the liver and activates the conversion of glucose to **glycogen** in its cells. This in turn tends to decrease blood glucose concentration and also creates and maintains a supply of glycogen stored in the liver cells which can release glucose as needed.

Hint

Chronic elevation means that the levels are excessively high over long periods of time.

Hint

Renal means related to kidney and kidney function.

Make the link

There is more about peripheral nerves in Chapter 3.1 on page 194.

Insulin

Hormone produced in the pancreas which stimulates the liver to convert glucose into glycogen.

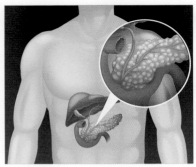

Figure 2.8.1 *The location of the liver and pancreas in the human body. The pancreas is also shown magnified and the liver is the large organ above it*

Between meals, pancreatic receptors respond to lowered blood glucose levels by increasing secretion of the hormone **glucagon** from pancreas cells. Glucagon activates the conversion of glycogen to glucose in the liver and so increases blood glucose concentration. Both insulin and glycogen work on **negative feedback** loops so that the action of each hormone is reduced as the glucose levels return to their normal ranges, as shown in **Figure 2.8.2**.

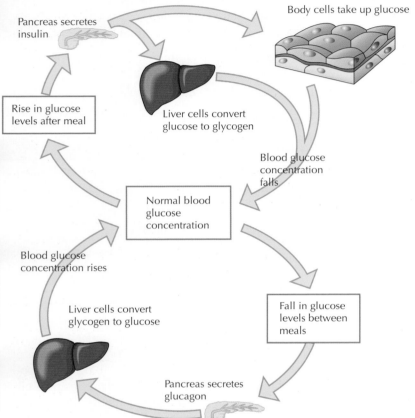

Figure 2.8.2 *The negative feedback loops involved in the regulation of blood glucose concentration*

📖 Glycogen

Storage carbohydrate produced by the liver from excess blood glucose.

📖 Glucagon

Hormone produced by the pancreas which stimulates the conversion of stored glycogen to glucose.

📖 Negative feedback

Method of control used in homeostasis.

📖 Adrenaline

Hormone that stimulates the release of glucose from glycogen during stress or high-intensity exercise.

Flight or fight

During exercise and especially during stress activities such as base-jumping (see **Figure 2.8.3**), the body might experience a so-called **adrenaline** rush. This is also sometimes called the fight or flight response. In response to the stimulus of a competitive, stressful or exciting situation, the hormone adrenaline is released from the adrenal glands into the bloodstream. It raises glucose concentrations in the blood by stimulating glucagon secretion and inhibiting insulin secretion. The idea is that the extra glucose is available to release energy to help the body cope with the flight or fight.

BLOOD GLUCOSE LEVELS AND OBESITY ●

Type 1 and type 2 diabetes

Type 1 diabetes

Type 1 diabetes usually occurs in childhood. A person with type 1 diabetes is unable to produce enough insulin. This inevitably means that their blood glucose levels rise quickly after a meal and remain high. They do not produce glycogen and excess glucose may be excreted in their urine.

The illness can be treated with regular doses of insulin which are usually taken by injection, as shown in **Figure 2.8.4**. The insulin can be obtained from animal sources, from genetically engineered bacteria or by laboratory synthesis from component amino acids.

Type 2 diabetes

Type 2 diabetes typically develops later in life. In type 2 diabetes, individuals produce insulin but their cells are less sensitive to it. This insulin resistance is linked to a decrease in the number of insulin receptors in the liver, leading to a failure to convert glucose to glycogen. Inevitably, glucose level rise after a meal and remain high. The likelihood of developing type 2 diabetes is increased by being overweight.

In both types of diabetes, individual blood glucose concentrations rise rapidly after a meal. The kidneys will remove some of this glucose, resulting in glucose appearing in urine. Testing for the presence of glucose in the urine is often used to indicate diabetes.

Glucose tolerance test

The **glucose tolerance test** is used to diagnose diabetes. The blood glucose concentrations of the individual taking the test are initially measured after fasting for at least eight hours. The individual then drinks a standard glucose solution and changes in their blood glucose concentration are measured for at least the next two hours. The blood glucose concentration of a diabetic usually starts at a higher level than that of a non-diabetic. During the test, a diabetic's blood glucose concentration increases to a much higher level than that of a non-diabetic and takes longer to return to its starting concentration, as shown in **Figure 2.8.5**.

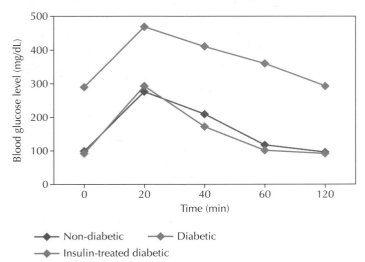

Figure 2.8.3 *This base-jumper may well experience an adrenaline rush as he takes the plunge*

📖 Type 1 diabetes

Condition in which an individual does not produce enough insulin.

Figure 2.8.4 *Type 1 diabetes is treated by regulated dosages of insulin usually taken by injection*

📖 Type 2 diabetes

Condition in which an individual's cells lose their sensitivity to insulin.

📖 Glucose tolerance test

Diagnostic test for diabetes in which blood glucose concentration is measured at time intervals after fasting and then after taking a standard glucose drink.

Figure 2.8.5 *Glucose tolerance curves for a non-diabetic, a diabetic and an insulin-treated diabetic*

Non-diabetic — Diabetic —
Insulin-treated diabetic —

Obesity and BMI

Obesity is characterised by excess body fat in relation to lean body tissue such as muscle.

Obesity may impair health and is a major risk factor for cardiovascular disease (CVD) and type 2 diabetes. **Body mass index (BMI)** is commonly used to measure obesity but can wrongly classify muscular individuals as obese. BMI is calculated using measured aspects of the body such as mass and height.

$$BMI = \frac{body\ mass\ (kg)}{height\ (m)^2}$$

Hint

In your exam, BMI should be calculated using scientific units: kilograms (kg) and metres (m).

On the NHS website, individuals are prompted to calculate BMI using non-scientific but widely used units: stones and pounds for mass and feet and inches for height. Note that the website provides buttons to convert to scientific units as indicated in Activity 2.8.2 Question 2 below.

A BMI greater than 30 on the international scale is used to indicate obesity, as shown in **Figure 2.8.6**.

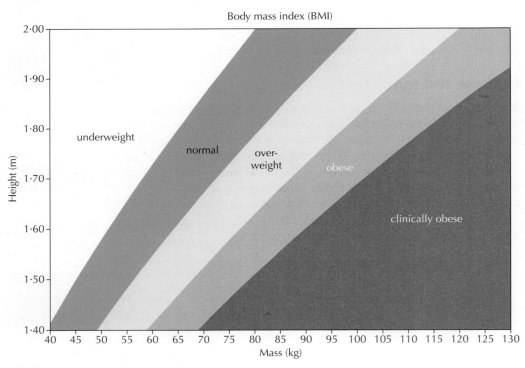

Figure 2.8.6 *BMI ranges and their classification*

Lifestyle choices are important considerations for reducing obesity, the levels of CVD and type 2 diabetes in the population. Obesity is linked to diets with high total fat and especially diets high in saturated fats. The energy intake in the diet should limit fats and free sugars. Fats have a high calorific value per gram and free sugars require no metabolic energy to be expended in their digestion.

Low levels of physical activity are also linked to obesity. Exercise increases energy expenditure and preserves lean tissue. Exercise can help to reduce risk factors for CVD by keeping weight under control, minimising stress, reducing hypertension and improving blood lipid profiles.

Make the link

There is more about hypertension in Chapter 2.6 on page 155.

Activity 2.8.1 Work individually to ...

Structured questions

1. **Name** the **three** hormones involved in the control of blood glucose concentrations. 3
2. **Name** the organ which contains receptor cells that can detect changes in blood glucose concentrations. 1
3. **Name** the organ in which excess glucose is converted to glycogen for storage. 1
4. **Describe** the differences between type 1 and type 2 diabetes. 2
5. **Give** the equation used to calculate body mass index (BMI). 1
6. **Describe** the limitations of using BMI to gauge obesity. 1
7. An individual with type 2 diabetes took a glucose tolerance test. At the same time doctors measured the insulin concentration in their blood. The results of these tests are shown in the graph.

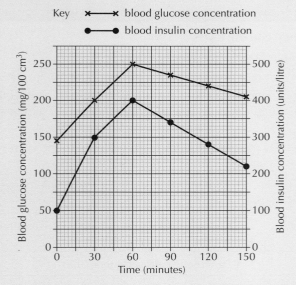

a) i. **Give** the insulin concentration when the blood glucose concentration was 250 mg/100 cm^3. 1
   ii. **Calculate** the percentage increase in blood insulin 60 minutes after taking the glucose drink. 1
   iii. The individual's blood volume was 4·8 litres.

   Calculate the total mass of glucose in their bloodstream 30 minutes after taking the glucose drink. 1

(continued)

    iv. **Calculate** the decrease in blood glucose concentration between 60 and 150 minutes after taking the glucose drink.   1

    v. **Predict** the blood glucose concentration of this individual 180 minutes after the start of the test.   1

b) i. **Describe** evidence from the graph which indicates that this individual does not have type 1 diabetes.   1

    ii. **Explain** why this individual cannot properly control their blood glucose concentrations.   1

> ### ⚬⚬ Make the link
>
> There is more about selecting information, processing information and predicting from a double axis line graph in Chapter 4 on page 267.

Extended response questions

1. **Write notes** on the use of body mass index (BMI) in the medical classification of body mass.   4

2. **Write notes** on the role of insulin in blood glucose regulation and the diagnosis and treatment of type 1 and type 2 diabetes.   10

GO! Activity 2.8.2 Work in pairs to …

1. **Card sequencing activity: Control of blood glucose levels**

   You will need: a stopwatch and a mini whiteboard with a marker pen.

   - Read the phrases in the grid below, which describe the control of blood glucose levels.
   - Work together to arrange the phrases in the correct order – start with phrase 5. Mark your answer on the whiteboard. Your teacher will check your work or put the correct sequence on the board.
   - You should each try this again individually against the clock – who is faster to the correct answer?
   - Your teacher may provide you with a photocopy of the grid. If so, cut it into strips, re-sequence it and glue it into your notes to make a permanent flowchart.

1 Glucose is absorbed by liver cells and converted to glycogen	**5** Glucose concentration in blood rises after a meal
2 Liver cells are activated by insulin	**6** Pancreatic cells secrete insulin
3 Insulin travels from the pancreas to the liver	**7** Glucose concentration in the bloodstream falls
4 Receptors in the pancreas respond to raised blood glucose concentration	**8** Negative feedback reduces insulin levels

2. **Research body mass index (BMI).**

   Visit the NHS web page below:

   www.nhs.uk/live-well/healthy-weight/bmi-calculator/

   You should each write a short report (100 words) which outlines the limitations of BMI.

Work out your own BMI if you want to – the NHS site asks you to use non-scientific measurements in stones and pounds and feet and inches but notice the buttons which allow you to convert your measurements to scientific units.

3. **Flashcard activity**

You will each need: a set of blank flashcards (A7 cards) and a stopwatch.

- Find the glossary terms for this chapter – they are the **black** typeface and **red** typeface terms. Using your blank cards, you should each make a set of flashcards for these terms – write the term on one side and the definition on the other. You will find the definitions in the chapter.
- Shuffle your cards and lay them out in a column, some showing terms and some showing definitions – you decide. Your partner should match their cards with yours, laying their cards in a column beside yours to give the corresponding term or definition. Time how long they take to do this.
- Now swap roles – your partner should lay out their cards and you should try the matching exercise while your partner times you.
- You should each keep your set of flashcards as a revision tool for later.

GO! Activity 2.8.3 Work as a group to ...

1. **Design and make a poster called Healthy Lifestyle.**

You will need: A3 paper or card and coloured pencils, or a mini whiteboard and marker pens.

Your poster must contain information which covers the success criteria 30–41 in the knowledge and understanding list at the end of this chapter.

Ensure that all the specialised glossary terms included in the criteria are mentioned.

Learning checklist

After working on this chapter, I can:

Knowledge and understanding

1. State that chronic elevated blood glucose levels lead to blood vessel damage and atherosclerosis.

2. State that chronic elevation of blood glucose levels due to untreated diabetes leads to the endothelial cells lining the blood vessels taking in more glucose than normal, which damages the blood vessels.

3. State that atherosclerosis may develop leading to cardiovascular disease (CVD), stroke or peripheral vascular disease which affects blood vessels leading to arms, hands, legs, feet and toes.

4. State that small blood vessels damaged by elevated glucose levels may result in haemorrhage of blood vessels in the retina, renal failure or peripheral nerve dysfunction.

181

5. State that blood glucose concentration is maintained within fine limits by negative feedback control involving the hormones insulin, glucagon and adrenaline.

6. State that blood glucose concentration is monitored by receptors in the pancreas.

7. State that the pancreas controls blood glucose with the two hormones insulin and glucagon, which act antagonistically: insulin decreasing blood glucose concentration and glucagon increasing it.

8. State that the hormones are transported in the blood to the liver.

9. State that pancreatic receptors respond to raised blood glucose levels by increasing secretion of insulin from the pancreas.

10. State that insulin activates the conversion of glucose to glycogen in the liver, decreasing blood glucose concentration.

11. State that pancreatic receptors respond to lowered blood glucose levels by increasing secretion of glucagon from the pancreas.

12. State that glucagon activates the conversion of glycogen to glucose in the liver, increasing blood glucose concentration.

13. State that during exercise and fight or flight responses, glucose concentrations in the blood are raised by adrenaline, released from the adrenal glands, stimulating glucagon secretion and inhibiting insulin secretion.

14. State that diabetics are unable to control their glucose concentration.

15. State that vascular disease can be a chronic complication of diabetes.

16. State that type 1 diabetes usually occurs in childhood.

17. State that a person with type 1 diabetes is unable to produce enough insulin and can be treated with regular doses of insulin.

18. State that type 2 diabetes typically develops later in life.

19. State that the likelihood of developing type 2 diabetes is increased by being overweight.

20. State that in type 2 diabetes, individuals produce insulin but their cells are less sensitive to it.

21. State that this insulin resistance is linked to a decrease in the number of insulin receptors in the liver, leading to a failure to convert glucose to glycogen.

22. State that in both types of diabetes, individual blood glucose concentrations will rise rapidly after a meal.

23. State that the kidneys will remove some of this glucose, resulting in glucose appearing in urine.

24. State that testing urine for glucose is often used as an indicator of diabetes.

25. State that the glucose tolerance test is used to diagnose diabetes.

26. State that the blood glucose concentrations of the individual are initially measured after fasting.

27. State that the individual then drinks a glucose solution and changes in their blood glucose concentration are measured for at least the next two hours.

28. State that the blood glucose concentration of a diabetic usually starts at a higher level than that of a non-diabetic.

29. State that during the test a diabetic's blood glucose concentration increases to a much higher level than that of a non-diabetic and takes longer to return to its starting concentration.

30. State that obesity may impair health.

31. State that obesity is a major risk factor for cardiovascular disease and type 2 diabetes.

32. State that obesity is characterised by excess body fat in relation to lean body tissue such as muscle.

33. State that body mass index (BMI) is a measurement of body fat based on height and weight.

34. State that BMI = body mass / height squared.

35. State that BMI can be used to indicate obesity, overweight, normal or underweight.

36. State that a BMI greater than 30 kg m^{-2} is used to indicate obesity.

37. State that body mass index (BMI) is commonly used to measure obesity but can wrongly classify muscular individuals as obese.

38. State that obesity is linked to high fat diets and a decrease in physical activity.

39. State that the energy intake in the diet should limit fats and free sugars, as fats have a high calorific value per gram and free sugars require no metabolic energy to be expended in their digestion.

183

40. State that exercise increases energy expenditure and preserves lean tissue.

41. State that exercise can help to reduce risk factors for cardiovascular disease (CVD) by keeping weight under control, minimising stress, reducing hypertension and improving blood lipid profiles (HDL:LDL).

Skills

1. *Select information from a double axis line graph.*

2. *Process information from a double axis line graph by calculations.*

3. *Draw conclusions from a double axis line graph.*

4. *Make predictions from a double axis line graph.*

Chapter 2 practice area test
Physiology and health

Write your answers on separate sheets of paper. Mark your work using the answers online at www.collins.co.uk/pages/Scottish-curriculum-free-resources.

Paper 1: Multiple choice

Total: 10 marks

1. A function of the germline cells in the testes is to produce

   A testosterone

   B sperm

   C seminal fluid

   D releaser hormone.

2. Changes in the ovary during the menstrual cycle are described below.

   1. Corpus luteum develops

   2. Ovulation occurs

   3. Progesterone is produced

   4. Corpus luteum degenerates

   5. Follicle matures

   The sequence in which these changes occur following menstruation is:

   A 2, 3, 1, 5, 4

   B 2, 1, 3, 4, 5

   C 5, 3, 2, 1, 4

   D 5, 2, 1, 3, 4

3. In the treatment of infertility, ovulation can be stimulated by using drugs that prevent the negative feedback effect of

   A oestrogen on FSH secretion

   B oestrogen on LH secretion

   C progesterone on FSH secretion

   D progesterone on LH secretion.

4. Phenylketonuria (PKU) is a metabolic disorder caused by a single autosomal gene.

   A man and a woman who are both unaffected have an affected child.

   What is the probability that their next child will be affected?

   A 25%

   B 50%

   C 75%

   D 100%

5. Which of these cross sections through a blood vessel represents an artery?

6. Which line in the table below describes the state of the heart valves during atrial and ventricular diastole?

	Valves	
	Atrio-ventricular (AV)	Semi-lunar (SL)
A	Open	Open
B	Closed	Closed
C	Open	Closed
D	Closed	Open

7. The graph shows how increasing oxygen uptake affects heart rate and stroke volume.

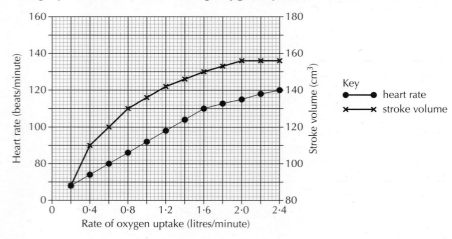

What is the cardiac output when the rate of oxygen uptake is 1·6 litres/minute?

A 1·36 cm³/min

B 176 cm³/min

C 15,225 cm³/min

D 16,500 cm³/min

8. Which of the following statements describes the role of lipoprotein in the transport and elimination of excess cholesterol?

    A Low-density lipoprotein (LDL) transports excess cholesterol from the liver to the body cells.

    B Low-density lipoprotein (LDL) transports excess cholesterol from the body cells to the liver.

    C High-density lipoprotein (HDL) transports excess cholesterol from the body cells to the liver.

    D High-density lipoprotein (HDL) transports excess cholesterol from the liver to the body cells.

9. The flow diagram shows how the concentration of glucose in the blood is controlled following a meal.

Blood glucose concentration increases

Increased secretion of hormone X by organ Y

Increased conversion of glucose to glycogen

Blood glucose concentration decreases

Which row in the table identifies hormone X and organ Y?

	Hormone X	*Organ Y*
A	Insulin	Liver
B	Glucagon	Liver
C	Insulin	Pancreas
D	Glucagon	Pancreas

10. By calculating body mass index (BMI), it can be determined whether a person is clinically obese.

The table below contains information about four individuals.

Individual	Height (m)	Mass (kg)
1	1·60	90
2	2·10	130
3	1·80	95
4	1·30	56

Which of these individuals would be classified as obese?

A 1 only

B 1 and 4

C 2 and 3

D All of them

Paper 2: Structured and extended response

Total: 40 marks

1. The diagram represents gamete production in part of a testis.

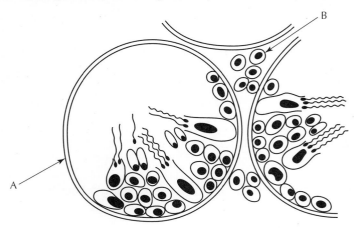

    a) **Name** structure A and **describe** its role in reproduction.   2

    b) **Name** a hormone produced by cell B.   1

2. The diagram shows how hormones from the pituitary gland affect the ovary.

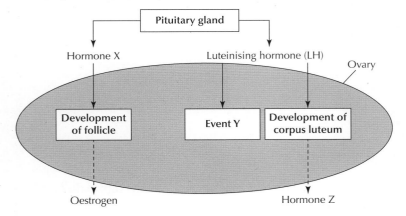

    a) i. **Name** hormones X and Z.   2

        ii. **Give** the term used to describe event Y.   1

    b) **Describe** the effects of oral contraceptives on the pituitary gland.   1

3. Duchenne muscular dystrophy is an inherited condition in which muscle fibres gradually degenerate. The condition is sex-linked and caused by a recessive allele.

The family tree shows the inheritance of the condition through three generations of a family.

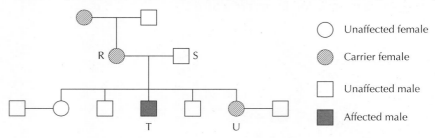

○ Unaffected female

◒ Carrier female

☐ Unaffected male

■ Affected male

a) Using the symbols D and d to represent the alleles, **state** the genotypes of individuals R and S. **2**

b) **Calculate** the percentage chance of any child of parents R and S having muscular dystrophy. **1**

c) **Calculate** the percentage chance of individual U and her partner having a son with muscular dystrophy. **1**

4. The diagram shows a capillary network with an area magnified to show the cells and vessels there.

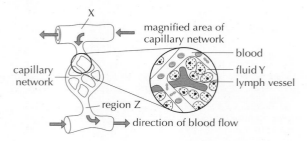

a) **Name** the type of blood vessel shown at X. **1**

b) **Describe** the change in blood pressure which would be expected as blood flowed from blood vessel X to region Z and **explain** why this change would occur. **2**

C) **Name** fluid Y. **1**

5. The diagram shows the heart and some of its associated nerves.

a) i. **Name** the region of the brain which regulates heart rate. **1**

ii. **Name** the antagonistic branches of the autonomic nervous system. **1**

b) i. **Name** region X. **1**

ii. **Describe** the role of nerve fibres Q in the cardiac cycle. **1**

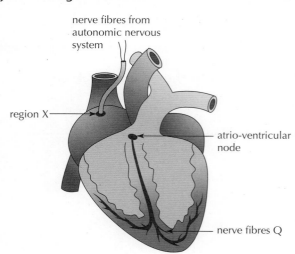

nerve fibres from autonomic nervous system

region X

atrio-ventricular node

nerve fibres Q

6. The diagram shows stages in a deep vein thrombosis (DVT), a blood clot that forms in a deep vein, in the lower leg.

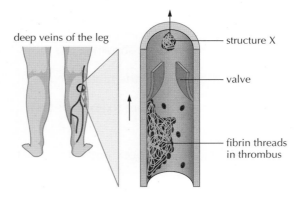

deep veins of the leg

structure X

valve

fibrin threads in thrombus

a) **Name** the enzyme that converts soluble fibrinogen into the threads of fibrin that form the thrombus. **1**

b) i. **Name** structure X that has formed by breaking free from the thrombus. **1**

   ii. **Name** the soluble protein present in blood plasma from which fibrin is produced. **1**

c) **Describe one** serious condition that could arise due to structure X. **1**

7. As part of a clinical trial, the systolic and diastolic blood pressures of eight adult volunteers were measured. Each volunteer was asked to drink 330 cm³ of a caffeine drink and their blood pressures were measured again one hour after taking the drink.

The results are shown in the table.

Volunteer	Initial blood pressure (mm Hg)		Blood pressure one hour later having ingested a caffeine drink (mm Hg)	
	Systolic	Diastolic	Systolic	Diastolic
1	130	84	139	90
2	126	75	133	83
3	127	80	144	88
4	129	72	135	80
5	134	81	143	87
6	119	72	127	82
7	123	73	133	82
8	120	71	126	80
Average	126	76	135	84

a) **Calculate** the percentage increase in the diastolic pressure of volunteer 3 one hour after taking the caffeine drink.   **1**

b) **State** the independent variable for this investigation.   **1**

c) **Draw a bar chart** to show all the average blood pressure readings shown in the table.   **2**

d) i. **Give one** conclusion which could be drawn from the results in the table.   **1**

   ii. **Suggest one** improvement that could be made to the procedure to improve the reliability of any conclusion drawn from the results.   **1**

8. The graph shows changes in the blood glucose concentrations in diabetic and non-diabetic individuals after each had consumed a glucose drink.

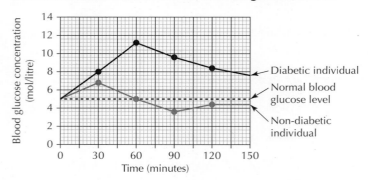

a) **Explain** why the blood glucose concentration in the non-diabetic individual increased more slowly than in the diabetic individual.   **2**

b) **Name** the hormone responsible for the increase in blood glucose concentration in the non-diabetic individual after 90 minutes.   **1**

9. **Discuss** the screening and testing procedures which may be carried out as part of antenatal care.   **8**

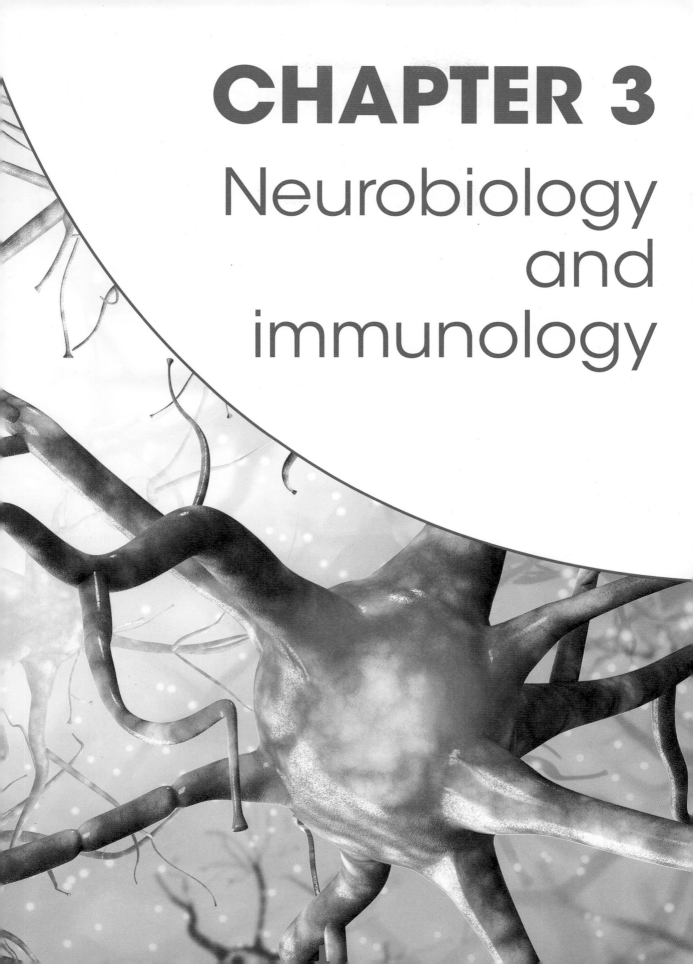

CHAPTER 3
Neurobiology and immunology

3.1 Divisions of the nervous system and neural pathways

You should already know:

- The nervous system consists of the central nervous system (CNS) and other nerves.
- The CNS consists of the brain and the spinal cord.
- The general structure and functions of the cerebrum, cerebellum and medulla.
- There are three types of neuron – sensory, inter and motor.
- Receptors detect sensory input/stimuli.
- Electrical impulses carry messages along neurons.
- Chemicals transfer these messages between neurons, at synapses.
- The structure and function of a reflex arc.
- A response to a stimulus can be a rapid action from a muscle or a slower response from a gland.
- Sensory neurons pass the information to the CNS.
- Inter neurons operate within the CNS and process information from the senses that requires a response.
- Motor neurons enable a response to occur at an effector (muscle or gland).
- Reflexes protect the body from harm.

Learning intentions

- Describe the structure and function of the central and peripheral nervous systems.
- Describe the function of the autonomic nervous system.
- Describe the structure and function of various neural pathways.

Divisions of the nervous system

The nervous system is a major body system in humans. The system detects change in the body's external and internal environments, processes that information and coordinates responses. It can be divided up in both structural and functional ways.

The structural divisions of the nervous system

📖 **Central nervous system (CNS)**

The brain and spinal cord.

Structurally, the nervous system can be thought of as the **central nervous system (CNS)** and the **peripheral nervous system (PNS)**. The CNS is made up of the brain and the spinal cord. The PNS consists of all of the nerves which lead into and out from the CNS as shown in **Figure 3.1.1**.

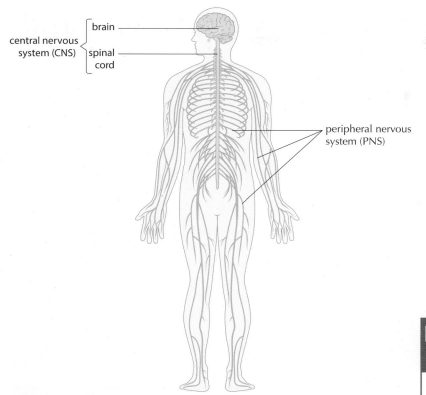

central nervous
system (CNS)

brain

spinal
cord

peripheral nervous
system (PNS)

Figure 3.1.1 *Structure of the nervous system*

The functional divisions of the nervous system

Functionally, the peripheral nervous system can be thought of as the **somatic nervous system (SNS)** and the **autonomic nervous system (ANS)**. The somatic nervous system controls voluntary actions. The autonomic nervous system controls actions which are not under voluntary control such as the heart rate, breathing rate, **peristalsis** and the flow of intestinal secretions. Peristalsis is the waves of muscular contractions which force food through the digestive system and intestinal secretions consist of the digestive juices containing the enzymes which break down the food molecules. The ANS consists of the **sympathetic** component which prepares the body for action and to cope with stress and the **parasympathetic** component which prepares the body for rest and recovery. **Figure 3.1.2** shows these functional divisions.

📖 **Somatic nervous system (SNS)**

The part of the peripheral nervous system which is under voluntary control.

📖 **Autonomic nervous system (ANS)**

The part of the peripheral nervous system which is under involuntary control.

📖 **Sympathetic**

Division of the autonomic nervous system which prepares the body for action.

📖 **Parasympathetic**

Division of the autonomic nervous system which prepares the body for rest and recovery.

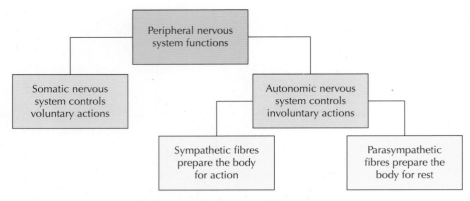

Figure 3.1.2 *The functional divisions of the peripheral nervous system*

Figure 3.1.3 *The accelerator and brake pedals in an automatic car are a bit like the sympathetic accelerator and the antagonistic parasympathetic brake – one is pressed as the other is released*

The actions of the sympathetic and parasympathetic systems are said to be **antagonistic** because they work together; as one system speeds up an action the other system slows it down – one acts as an accelerator and the other acts as a brake.

The sympathetic system prepares the body for action so speeds up heart rate and breathing rate while slowing down peristalsis and production of intestinal secretions. The parasympathetic system prepares the body for rest and recovery and so changes these in the opposite and antagonistic way as shown in the table and **Figure 3.1.3**.

Process	Sympathetic effects	Parasympathetic effects
Heart rate	Increases rate	Decreases rate
Breathing rate	Increases rate	Decreases rate
Peristalsis	Decreases rate	Increases rate
Release of intestinal secretions	Decreases rate	Increases rate

Neurons and neural pathways

Neurons are nerve cells. They are linked together to make flexible neural pathways to coordinate and integrate the actions of the nervous system.

The somatic nervous system contains sensory and motor neurons. Sensory neurons take impulses from receptors in sense organs to the CNS where they are processed. Motor neurons take impulses from the CNS to the effector muscles and glands where responses occur.

Converging neural pathways

In a **converging neural pathway**, impulses from several neurons travel towards and pass on impulses to a single neuron. This can increase sensitivity to excitatory or inhibitory signals. For example, the very many receptor cells in the retina of the eye generate nervous impulses which pass down many neurons and converge together to allow the integrated images of vision to be formed. In dim light the impulses are summated together to ensure image formation as shown in **Figure 3.1.4**.

Make the link

There is more about the detailed structure and functioning of neurons in Chapter 3.4 on page 214.

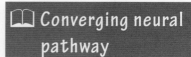

Converging neural pathway

Group of neurons which join to an increasingly small number of neurons which affect the sensory area of the cortex.

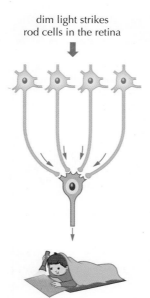

dim light strikes
rod cells in the retina

each rod cell generates an impulse
that is not strong enough to trigger
an impulse in the next neuron

impulses from several rod cells are
summated which allows an impulse
to be triggered in the next neuron
and on to the CNS

visual area of the brain
stimulated to allow some
sensitivity of the eye in
dim light conditions

Figure 3.1.4 *A converging neural pathway which allows the eye to be sensitive in dim light*

Diverging neural pathways

In a **diverging neural pathway**, impulses from a single neuron travel to several neurons and so affect more than one effector at the same time. For example, in the coordinated movements needed to write, many effector muscles can be activated from nervous impulses from the same source as shown in **Figure 3.1.5**.

Reverberating neural pathways

In a **reverberating neural pathway**, neurons later in the pathway link with earlier neurons, sending the impulse back through the pathway repeatedly. This allows the repeated stimulation of the same pathway needed to perform repetitive actions such as breathing as shown in **Figure 3.1.6**.

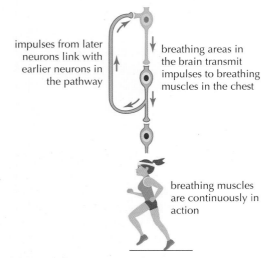

impulses from later
neurons link with
earlier neurons in
the pathway

breathing areas in
the brain transmit
impulses to breathing
muscles in the chest

breathing muscles
are continuously in
action

Figure 3.1.6 *A reverberating neural pathway which allows continuous repetition of the breathing actions*

> **Make the link**
>
> There is more about how converging pathways work in Chapter 3.4 on page 217.

> **Diverging neural pathway**
>
> Groups of neurons which split from each other and each of which affects a different effector.

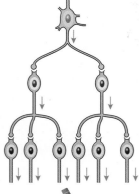

neuron in motor area
of cortex in brain generates
impulses to writing action

impulses diverge to many
neurons, allowing the
coordinated action of
many muscles

many muscles act
in coordination to
allow handwriting

Figure 3.1.5 *A diverging neural pathway which allows the coordinated muscle movements needed for writing*

> **Reverberating neural pathway**
>
> Organisation of neurons which loop back to give repeated stimulation of a single pathway.

> **Make the link**
>
> There is more about the autonomic control of the heart rate in Chapter 2.6 on page 156.

GO! Activity 3.1.1 Work individually to ...

Structured questions

1. **Name** the components of the central nervous system.   2
2. **Describe** the differences between somatic and autonomic actions in the body.   2
3. **Give two** examples of the effects of the parasympathetic nerves on the body.   2
4. As part of a clinical trial of two drugs for use in the treatment of certain heart conditions, a healthy male participant was asked to lie down and completely relax to allow his heart rate to come to a resting value. He was attached to a monitor to measure his heart rate continuously throughout the trial.

   After 30 seconds, he was given an injection of a sympathetic heart nerve blocker drug and after a further 30 seconds an injection of a parasympathetic heart nerve blocker was given. Each drug was expected to have effects up to 60 seconds following the injection.

   The results of the trial are shown in the graph.

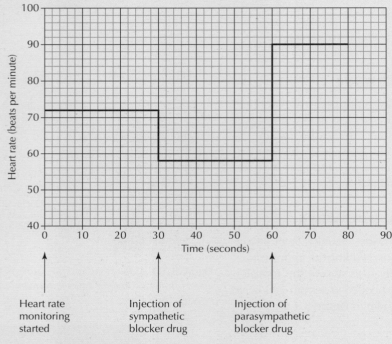

   a) **Give** the participant's resting heart rate.   1
   b) **Calculate** the percentage decrease in the heart rate as a result of the injection of the sympathetic blocker drug.   1
   c) **State** the increase in heart rate after the parasympathetic drug was injected.   1
   d) It was concluded that the parasympathetic nerve has a greater influence on the participant's heart rate than the sympathetic nerve.

      Describe evidence in the graph which supports this statement.   1
   e) **Identify** the nerve(s) which would be blocked 70 seconds into the trial.   1
   f) **Predict** the expected heart rate of the participant at 90 seconds.   1
   g) **Explain** why the results of the trial could not be considered reliable.   1

 **Make the link**

There is information about clinical trials in Chapter 3.8 on page 247.

 **Make the link**

There is information about selecting and processing information, concluding and predicting from line graphs and evaluation in Chapter 4 on page 267.

Extended response questions

1. **Write notes** on the effects of the sympathetic nervous system on the body. 4
2. **Give an account** of the different types of neural pathway. 9

Activity 3.1.2 Work in pairs to …

 1. **Research autonomic drugs.**

You should each write a 100 - word report on autonomic drugs.

Include an explanation of the term 'autonomic drug' and give a few examples of autonomic drugs and their uses.

 2. **Flashcard activity**

You will each need: a set of blank flashcards (A7 cards) and a stopwatch.

- Find the glossary terms for this chapter – they are the **black** typeface and **red** typeface terms. Using your blank cards, you should each make a set of flashcards for these terms – write the term on one side and the definition on the other. You will find the definitions in the chapter.

- Shuffle your cards and lay them out in a column, some showing terms and some showing definitions – you decide. Your partner should match their cards with yours, laying their cards in a column beside yours to give the corresponding term or definition. Time how long they take to do this.

- Now swap roles – your partner should lay out their cards and you should try the matching exercise while your partner times you.

- You should each keep your set of flashcards as a revision tool for later.

Activity 3.1.3 Work as a group to …

1. **Make models of neural pathways.**

You will need: a ball of string, Blu-Tack and scissors.

Work together to make models of diverging and converging neural pathways.

Use about 10–12 pieces of string to represent neurons on each pathway. Lay the pieces of string on the desk and fray one end to make 'connections'. Link the neurons together using Blu-Tack.

Be ready to describe your model of each pathway and have two examples of how each pathway works in the human body.

Learning checklist

After working on this chapter, I can:

Knowledge and understanding

1. State that the central nervous system (CNS) consists of the brain and the spinal cord.

2. State that the peripheral nervous system (PNS) consists of the somatic nervous system (SNS) and the autonomic nervous system (ANS).

3. State that the SNS controls voluntary actions by skeletal muscles.

4. State that the somatic nervous system contains sensory and motor neurons.

5. State that sensory neurons carry impulses from sense organs to the CNS.

6. State that motor neurons carry impulses from the CNS to muscles and glands.

7. State that the ANS controls involuntary actions by glands, smooth muscle and cardiac muscle.

8. State that the autonomic nervous system (ANS) consists of the sympathetic and parasympathetic systems, which are antagonistic to each other.

9. State that the sympathetic system speeds up heart rate and breathing rate while slowing down peristalsis and production of intestinal secretions.

10. State that the parasympathetic system slows the heart rate and breathing rate but speeds up digestive processes such as intestinal secretions and peristalsis.

11. State that in a converging neural pathway, impulses from several neurons travel to a single neuron. This increases the sensitivity to excitatory or inhibitory signals.

12. State that in a diverging neural pathway, impulses from one neuron travel to several neurons and so affect more than one destination at the same time.

13. State that in a reverberating neural pathway, neurons later in the pathway link with earlier neurons, sending the impulse back through the pathway. This allows repeated stimulation of the pathway.

Skills

1. *Select information from a line graph.*

2. *Process information selected from a line graph.*

3. *Draw conclusions from a line graph.*

4. *Make predictions by extrapolating from a line graph.*

5. *Evaluate the reliability of procedures.*

3.2 The cerebral cortex

You should already know:

- The location and functions of the cerebrum.

Learning intentions

- Describe the structure and function of the cerebral cortex.
- Describe the link between the cerebral hemispheres.

The cerebral cortex

The central nervous system (CNS) consists of the brain and the spinal cord. The brain is the central processing area for the nervous system and is divided into regions of specific function as shown in **Figure 3.2.1**.

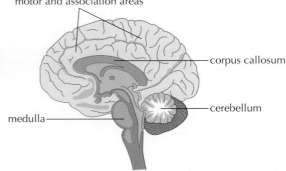

cerebral cortex containing sensory, motor and association areas

corpus callosum

cerebellum

medulla

Figure 3.2.1 *The human brain showing relatively large size of the cerebrum compared with the other regions of the brain*

The **cerebrum** is the large structure at the top and front of the brain. The functioning of the human cerebrum is a major difference between humans and their closest primate relatives such as gorillas and chimpanzees. The folded outer layer of the cerebrum is called the **cerebral cortex** and is packed with interconnected neurons. This region is the centre of our conscious thought, recalls memories and alters behaviour in the light of experience in the process of learning.

There is **localisation of brain functions** in the cerebral cortex. Different areas have different functions. There are **sensory areas**, **motor areas** and **association areas**. The sensory areas receive nervous impulses from the sensory **receptors** around the body and give a person information about stimuli which originate in the environment. The brain can process the incoming information and direct appropriate responses. Motor areas send nervous impulses out from the brain to the **effectors**, which are the muscles and glands. These bring about the appropriate responses. Much of the cortex is made up of association

Hint

The cerebellum is involved with balance and coordination although you are not required to know about this structure for your exam.

Make the link

There is more about the medulla, which controls autonomic responses such as heart rate, in Chapter 2.6 on page 156.

Cerebral cortex

Outer layer of the cerebrum, the site of conscious thought, memory and learning.

Localisation of brain functions

Concept which allocates a specific function to a specific area of the cerebral cortex.

Effectors

Muscles and glands which bring about responses to stimuli.

areas which are involved in specialised functions such as language processing, personality, imagination and intelligence.

Transfer of information between cerebral hemispheres

The cerebrum is divided into two separate halves called **cerebral hemispheres**. The left cerebral hemisphere deals with information from the right visual field and controls the right side of the body and the right cerebral hemisphere deals with information from the left visual field and controls the left side of the body. Transfer of information between the two cerebral hemispheres occurs through a band of nervous tissue which forms a bridge between them. The bridge is called the **corpus callosum** as shown in **Figure 3.2.2** and allows overall integration of brain functions.

Split-brain experiments

Much detailed information about the functioning of the corpus callosum has come from work with split-brain patients whose corpus callosum was surgically severed to reduce the symptoms and effects of conditions such as severe epilepsy. One important feature to note is that the association area for language is in the left cerebral hemisphere. This means that a split-brain patient will not be able to name objects which are in their left visual field. This is because the information about these objects is perceived on the right side of the brain but cannot cross to the left side to integrate with the language centre because of the severed corpus callosum as shown in **Figure 3.2.3**.

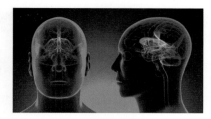

Figure 3.2.2 *The position of the corpus callosum, shown in orange, deep within the brain. It connects the two cerebral hemispheres, allowing information to be shared between them*

🔍 **Hint**

You don't have to know where the different areas of the cortex are located, just some examples of what these areas do.

📖 *Corpus callosum*

A band of tissue which connects the left and right cerebral hemispheres and allows integration of functions between the two.

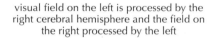

visual field on the left is processed by the right cerebral hemisphere and the field on the right processed by the left

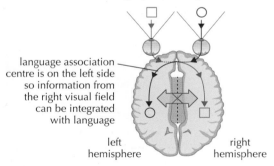

language association centre is on the left side so information from the right visual field can be integrated with language

left hemisphere | right hemisphere

Figure 3.2.3 *This split-brain patient sees the square on the left of their visual field and it is perceived by the right side of the brain because the optic cross over is not split. Because the corpus callosum is severed, information about the square cannot be integrated with the language area of the left side – they see it but they cannot name it!*

Early experiments with split-brain patients in the 1950s and 1960s, such as the one shown in **Figure 3.2.4**, were carried out and the results revealed much about the role of the corpus callosum in the transfer of information between cerebral hemispheres. Split-brain procedures were used in the 1950s and 1960s but are seldom used now and have been replaced by alternative treatments.

Left panel: The patient is asked to focus on the dot in the middle of the screen. A word is briefly flashed onto the right field. The left hemisphere processes the image and integrates with the language association area and the patient is able to say 'face'

Right panel: The word is now briefly flashed onto the left field. The right hemisphere processes the image but can't integrate the information with the language area on the left so the patient can't say what he sees – they can draw it with their left hand though because the right hemisphere controls the muscles on the left

Figure 3.2.4 *Split-brain experiment*

GO! Activity 3.2.1 Work individually to ...

Structured questions

1. **Give two** functions of the cerebral cortex.　　　　　　　　　　　　2

2. **Name** the **three** areas of localisation of function in the cerebral cortex.　　3

3. **Name** the band of tissue which connects the two cerebral hemispheres.　　1

4. An experiment was carried out to give information about the role of the corpus callosum in brain function. A volunteer split-brain patient was asked to focus on a spot in the middle of a screen. The words 'key' and 'spoon' were flashed briefly onto the screen in the positions shown in the diagram. The patient was then asked to use their left hand to pick up the objects they saw named on the screen.

   a) **Explain** why the patient picked up the key and did not pick up the spoon.　　2

   b) The patient was then asked to say what they saw written on the screen.

   Predict the answer they would have given and **state** the reason for your prediction.　　2

   c) **Suggest** a procedure which should be carried out to provide a control for this experiment.　　1

   d) **Suggest** what should be done to ensure that any conclusions made about the role of the corpus callosum in brain function are reliable.　　1

Make the link

There is information about controls, reliability and predictions from experiments in Chapter 4 on page 263.

Extended response question

1. **Give an account** of localisation of functions in the human cerebral cortex and the role of the corpus callosum in brain function. 10

 ## Activity 3.2.2 Work in pairs to …

1. **Flashcard activity**

   You will each need: a set of blank flashcards (A7 cards) and a stopwatch.

   • Find the glossary terms for this chapter – they are the **black** typeface and **red** typeface terms. Using your blank cards, you should each make a set of flashcards for these terms – write the term on one side and the definition on the other. You will find the definitions in the chapter.

   • Shuffle your cards and lay them out in a column, some showing terms and some showing definitions – you decide. Your partner should match their cards with yours, laying their cards in a column beside yours to give the corresponding term or definition. Time how long they take to do this.

   • Now swap roles – your partner should lay out their cards and you should try the matching exercise while your partner times you.

   • You should each keep your set of flashcards as a revision tool for later.

 ## Activity 3.2.3 Work as a group to …

1. **Research the work of the Nobel Prize winner Roger Sperry.**

   Visit the web page below:

   https://embryo.asu.edu/pages/roger-sperrys-split-brain-experiments-1959-1968

   On your own, consider the split-brain work of Roger Sperry, who won the Nobel Prize for Physiology or Medicine in 1981.

   As a group, collate your findings and make a three-slide PowerPoint presentation. The presentation should include:

   • a photograph of Roger Sperry

   • an explanation of a split-brain and why these patients had their brains surgically spilt

   • a summary of what Sperry discovered from his research

2. **Research split-brain studies.**

   Visit the web page below:

   www.nature.com/news/the-split-brain-a-tale-of-two-halves-1.10213

   On your own, consider the material in the article relating to Vicki – her story is covered in the first five paragraphs and under the sub-heading 'Subject of interest' later in the article.

   As a group, collate your findings and make a three-slide PowerPoint presentation entitled 'Vicki's Story'.

Learning checklist

After working on this chapter, I can:

Knowledge and understanding

1. State that the cerebral cortex is the centre of conscious thought. It also recalls memories and alters behaviour in the light of experience.

2. State that there is localisation of brain functions in the cerebral cortex. It contains sensory areas, motor areas and association areas.

3. State that there are association areas involved in language processing, personality, imagination and intelligence.

4. State that information from one side of the body is processed in the opposite side of the cerebrum.

5. State that the left cerebral hemisphere deals with information from the right visual field and controls the right side of the body and vice versa.

6. State that transfer of information between the cerebral hemispheres occurs through the corpus callosum.

Skills

1. *Describe a suitable control for a split-brain experiment.*

2. *Make predictions from a split-brain experiment.*

3. *Evaluate an experiment in terms of reliability.*

3.3 Memory

You should already know:

- The general structure and functions of the cerebrum.

Learning intentions

- Describe encoding, storage and retrieval of memories.
- Describe sensory memory.
- Explain the role and characteristics of the short-term memory (STM).
- Explain the role and characteristics of the long-term memory (LTM).

Memory

Memory is a crucial part of the functioning of the cerebrum. Memories include past experiences, knowledge and thoughts. Memory function involves processes called **encoding**, **storage** and **retrieval** of information as shown in **Figure 3.3.1**.

Figure 3.3.1 *The processes involved in memory*

Encoding involves the processing of information in preparation for storage. Storage is the maintenance of information over periods of time and retrieval is the finding and recalling of information from the store.

Sensory memory

All the information which enters the brain passes through a phase called **sensory memory**. The sensory memory deals with environmental stimuli such as the light involved with visual input and sound involved with auditory input. The sensory memory only lasts for a few seconds, during which time a selected proportion of the stimuli can be encoded into the next phase called **short-term memory (STM)**.

Short-term memory (STM)

Short-term memory (STM) has a short span with a limited capacity and the ability to hold information only for a short time. Experiments carried out in the 1950s suggest that the number of items in an average human **memory span** is between 5 and 9, which is sometimes referred to as Miller's Law.

The capacity of STM can be improved by **chunking**. Chunking is where items are grouped together so that they are stored as a single item in the STM. For example, when trying to quickly memorise a sequence of digits, say the number of a road, the numbers can be

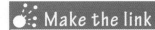
Make the link

There is more about the cerebrum in Chapter 3.2 on page 202.

📖 Sensory memory

Phase of memory into which sensory information is initially and very briefly stored.

📖 Short-term memory (STM)

Phase of memory into which selected information from sensory memory is stored.

📖 Chunking

Arranging items into groups to improve the capacity of STM.

📖 **Working memory model**

Model of memory which takes account of the ability to process items within the STM.

📖 **Serial position effect**

The pattern of recall of items from a sequentially presented list.

mentally grouped so that the B2375 can be thought of as the B twenty-three (first item), seventy-five (second item).

The STM can perform a limited amount of processing as well as storing so that simple cognitive tasks can be performed within it even though the items may eventually be discarded. This model of memory with processing being performed within the STM is known as the **working memory model**.

The serial position effect

The **serial position effect** can be seen in the results of experiments designed to measure the capacity of STM span. If a subject is given a rapidly presented sequence of items and asked to recall these items in order soon after, the serial position effect can be noted. As the first few items are presented, the subject tries to maintain these items by rehearsal while the middle items are being presented. This tends to limit the recall of the middle items which are displaced by the last few items presented and are lost or their memory decays. The last few items are still in the STM when recall is asked for.

Figure 3.3.2 shows the general results of a serial position experiment. The recall of the items presented first is called **primacy** and of the last few items is **recency**. Intermediate items are lost by **displacement** or decay. Recency can disappear if the time gap between the presentation of the list of items and recalling the list is extended.

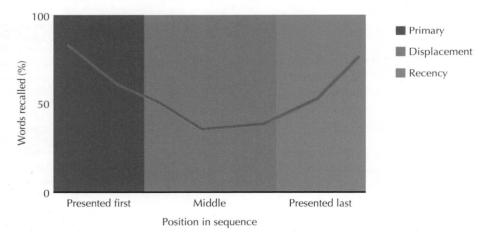

Figure 3.3.2 *The serial position effect – note the primacy, displacement and recency effects*

Long-term memory (LTM)

📖 **Long-term memory**

Phase of memory into which selected information from STM is permanently stored.

Long-term memory refers to memories held indefinitely over long periods of time. LTM seems to have an unlimited capacity and holds information potentially over a lifetime. The information in the LTM has to be transferred to it through the STM. This transfer is natural but is stimulated by the processes of **rehearsal**, **organisation** and **elaboration**. Rehearsal involves information being repeated or practised to encourage its entry to LTM. This method of encoding can be described as shallow as it may produce memories which are not so robust and are less easily retrieved. Organisation involves manipulating, classifying, grouping or sequencing information to increase the chances of transfer and is regarded as a deeper form of encoding when compared to rehearsal. The deepest encoding is elaboration, in which support information is added to improve its retention in the LTM. An example of the increasing depth of encoding is shown in **Figure 3.3.3**.

Rehearsal – e.g. trying to remember the colours of the visible spectrum seen in a rainbow by rehearsing a list	**Organisation** – e.g. trying to remember the colours of the visible spectrum seen in a rainbow by putting them into order of wavelength	**Elaboration** – e.g. trying to remember the colours of the visible spectrum seen in a rainbow by putting them into order and adding information as a mnemonic
Blue, red, orange, green, indigo, yellow, violet	Red, orange, yellow, green, blue, indigo, violet or ROY – G – BIV	**R**ed, **o**range, **y**ellow, **g**reen, **b**lue, **i**ndigo, **v**iolet '**R**ichard **of Y**ork **g**ave **b**attle **in v**ain' – a first letter mnemonic (nothing to do with the rainbow!)

Increasing depth of encoding →

Figure 3.3.3 *Can you remember the colours of the rainbow? Increasing the depth of encoding helps*

Contextual cues

Contextual cues help to retrieve items from the LTM. They relate to the time and place when and where the information was initially encoded into LTM and release memories which might otherwise be impossible to retrieve. Contextual cues are used by psychiatrists trying to probe into a patient's past to retrieve memories which the patient may well be trying to block. Police investigators use contextual cues by setting a reconstruction of a crime to try to jog the LTM of potential witnesses who may not even realise they have these memories. Many people keep old photographs to remind them of times and places in the past.

📖 **Contextual cue**

A clue or trigger to the time and place when and where a memory was encoded.

Figure 3.3.4 *(a) A reconstruction of a crime scene uses contextual cues to try to jog the memory of potential witnesses (b) A psychiatrist may use questions with contextual cues to try to unlock memories in a patient's long-term memory (c) Old photographs act as contextual cues to memories of the past*

GO! Activity 3.3.1 Work individually to ...

Structured questions

1. **State** the meaning of the term 'sensory memory'.   1
2. **Give two** characteristics of short-term memory (STM).   2
3. **Describe** what is meant by 'chunking'.   2
4. **Explain** the serial position effect.   3
5. **Give three** methods of encouraging the transfer of information to long-term memory (LTM).   3
6. An investigation was carried out into the serial position effect. A sequence of 10 different named pictures was shown one at a time to five 12-year-old children. After a gap of 1 minute, the children were asked to write down the names of the pictures they saw in the order that they appeared. The results were analysed and are shown in the table.

Child	Position of the picture in the presentation sequence									
	1	2	3	4	5	6	7	8	9	10
1	✓	✓	✓			✓			✓	✓
2	✓	✓	✓	✓	✓			✓	✓	✓
3	✓	✓		✓	✓				✓	✓
4	✓		✓	✓			✓	✓		✓
5	✓	✓	✓					✓	✓	✓
Percentage correct recall (%)	100	80	80	60	40	20	20	60	80	100

a) **Draw a line graph** to show the percentage of children who recalled the pictures in each position in the presentation sequence correctly.   2

b) **Give two** variables which should be kept the same for each of the images in this investigation.   2

c) **Give two** conclusions which can be drawn from the results of this investigation.   2

d) **Suggest** how the reliability of the results could have been improved.   1

e) **Predict** how the results would be different if the time between the presentation and the writing down of the responses had been made longer.   1

Make the link

There is information about presenting, processing, concluding, predicting and evaluating in Chapter 4 on page 263.

Extended response questions

1. **Write notes on** short-term memory (STM).   4
2. **Give an account of** memory under the following headings:
   a) Encoding   2
   b) Transfer from STM to LTM   3
   c) Retrieval from long-term memory (LTM)   2

Activity 3.3.2 Work in pairs to ...

1. **Card sequencing activity: Events in memory**

   You will need: a stopwatch and a mini whiteboard with a marker pen.

   - Read the phrases in the grid below, which describe events in memory.
   - Work together to arrange the phrases in the correct order – start with phrase 5. Mark your answer on the whiteboard. Your teacher will check your work or put the correct sequence on the board.
   - You should each try this again individually against the clock – who is faster to the correct answer?
   - Your teacher may provide you with a photocopy of the grid. If so, cut it into strips, re-sequence it and glue it into your notes to make a permanent flowchart.

1 Selected auditory and visual information passes to short-term memory (STM)
2 Short-term memory (STM) stores information briefly
3 Selected items of information transferred to long-term memory (LTM)
4 Information encoded for transfer to long-term memory (LTM)
5 Information enters sensory memory
6 Information stored in long-term memory (LTM) for long periods
7 Information retrieved from long-term memory (LTM)

 2. **Flashcard activity**

   You will each need: a set of blank flashcards (A7 cards) and a stopwatch.

   - Find the glossary terms for this chapter – they are the **black** typeface and **red** typeface terms. Using your blank cards, you should each make a set of flashcards for these terms – write the term on one side and the definition on the other. You will find the definitions in the chapter.
   - Shuffle your cards and lay them out in a column, some showing terms and some showing definitions – you decide. Your partner should match their cards with yours, laying their cards in a column beside yours to give the corresponding term or definition. Time how long they take to do this.
   - Now swap roles – your partner should lay out their cards and you should try the matching exercise while your partner times you.
   - You should each keep your set of flashcards as a revision tool for later.

 Activity 3.3.3 Work as a group to ...

1. **Practical activity: serial position effect**

   You will need: a set of 12 large cards each with a random letter printed on it or a named picture of an object pasted onto it – your teacher may supply these, a stopwatch and a supply of pieces of A6 paper.

   Method

   - Two classmates should be selected as experimenters and the rest as the subjects. Each of the subjects should be issued with a piece of paper.

(Continued)

- The experimenters should put the large cards into random order and each record the order. The experimenters should then sit behind a desk in front of the class.
- The experimenters lift each card in turn to show to the class, concealing the card again after 2 seconds exposure.
- Once all of the cards have been shown, the subjects are given 1 minute to write down the letters/names which they can recall and the order in which they were shown.
- The papers are taken in after 1 minute and the experimenters then work out the number of correct answers for each position in the sequence. They record the result and display them on a screen or on the board for everyone to see.

Each student should write up a short report of the experiment, including the method, a results table, a graph of the percentage recall of items against their position in the sequence and a conclusion.

Assignment Support

You could use this experimental technique to generate data for your assignment:

- You could investigate many aspects of the serial position effect, such as the effect of delay time before recall, the effect of time between presentation of items, the effect of distraction on recall and many others.
- You could use a copy of the grid on page 288 to plan an assignment based on this experiment.
- An exemplar of an assignment on this topic can be found on the SQA Understanding Standards website at:

  www.understandingstandards.org.uk/Subjects/HumanBiology/Assignment

- You should read the commentary carefully and note the reasons why certain marks were not awarded.

Learning checklist

After working on this chapter, I can:

Knowledge and understanding

1. State that memories include past experiences, knowledge and thoughts.

2. State that memory involves encoding, storage and retrieval of information.

3. State that all information entering the brain passes through sensory memory and enters short-term memory (STM). Information is then either transferred to long-term memory (LTM) or is discarded.

4. State that sensory memory retains all the visual and auditory input received for a few seconds. Only selected images and sounds are encoded into short-term memory (STM).

5. State that STM has a limited capacity and holds information for a short time.

6. State that the capacity of STM can be improved by 'chunking'.

7. State that STM can also process data, to a limited extent, as well as store it.

8. State that this working memory model explains why the STM can perform simple cognitive tasks.

9. State that memory span is the number of discrete items, such as letters, words or numbers, that the STM can hold.

10. State that the serial position effect is the tendency of a person to recall the first and last items in a series best, and the middle items worst.

11. State that items can be retained in the STM by rehearsal or lost by displacement or decay.

12. State that LTM has an unlimited capacity and holds information for a long time.

13. State that information can be transferred from STM to LTM by rehearsal, organisation and elaboration.

14. State that information can be encoded by shallow processing or deep (elaborative) processing.

15. State that rehearsal is regarded as a shallow form of encoding information into LTM.

16. State that elaboration is regarded as a deeper form of encoding which leads to improved information retention.

17. State that retrieval from LTM is aided by the use of contextual cues.

18. State that contextual cues relate to the time and place when and where the information was initially encoded into LTM.

Skills

1. *Present information as a line graph.*

2. *Plan a procedure through identification of variables.*

3. *Evaluate a procedure in terms of reliability.*

4. *Make predictions.*

5. *Process information by calculating percentages.*

3.4 The cells of the nervous system and neurotransmitters at synapses

You should already know:

- Sensory neurons pass information to the central nervous system (CNS).
- Inter neurons operate within the CNS and process information from the senses that requires a response.
- Motor neurons enable a response to occur at an effector, which can be a muscle or a gland.
- Receptors detect stimuli.
- Electrical impulses carry messages along neurons.
- Chemicals transfer these messages between neurons, at synapses.

Learning intentions

- Describe the structure and function of neurons.
- Explain the role of glial cells in the nervous system.
- Describe the role of neurotransmitters in the transmission of nerve impulses at synapses.
- Describe the effect of various neurotransmitters and drugs on mood and behaviour.

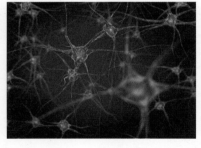

Figure 3.4.1 *Visualisation of brain cells teased apart*

📖 Dendrite

Short nerve fibre which conducts nervous impulses towards the cell body of a neuron.

📖 Axon

Long nerve fibre which conducts nervous impulses away from the cell body of a neuron.

Structure and function of cells in the nervous system

The nervous system is a hugely complex network of interconnected fibrous cells which work together in an integrated way to control and coordinate the body. **Figure 3.4.1** illustrates a tiny sample of brain cells to show their fibres and connections.

Neurons

Neurons are the commonest type of cell in the nervous system. Like most human cells, they each have a nucleus and cytoplasm and are bounded by a cell membrane. Neurons have their cytoplasm drawn out into a long fibre or fibres which can conduct electrical impulses. The region of a neuron with its nucleus and most of the cytoplasm is called the **cell body** and the fibres are named according to the direction of flow of the electrical impulses through them. **Dendrites** carry impulses towards the cell body of a neuron and **axons** carry them away. **Figure 3.4.2** shows a sensory and a motor neuron.

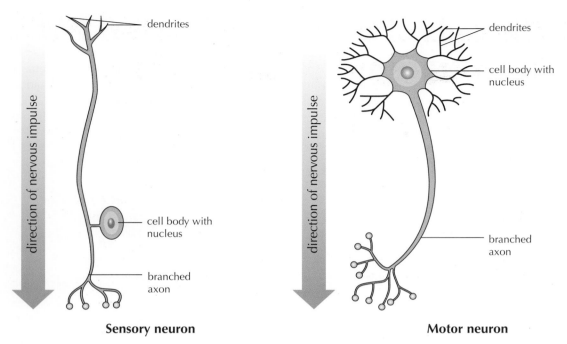

Sensory neuron　　　　　　　**Motor neuron**

Figure 3.4.2 *Basic structure of sensory and motor neurons – notice the direction of flow of electrical impulses in dendrites towards the cell body and away from the cell body through axons*

Long fibres such as axons are surrounded by a sheath made of a fatty substance called **myelin**, which insulates the axon and increases the speed of impulse conduction. **Myelination** continues from birth into adolescence. Responses to stimuli in the first few years of life are not as rapid or coordinated as those of an older child or adult. **Figure 3.4.3** shows the myelin sheath round a motor neuron.

📖 Myelination

Process by which additional myelin is added round the long fibres of a neuron.

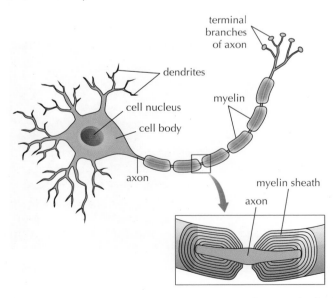

Figure 3.4.3 *Myelin sheath of a motor neuron showing the build-up of layers of fatty myelin*

Hint

You don't need to know about MS specifically for your exam but you do need to know about autoimmune disorders in general.

Make the link

There is more about autoimmune disorders in Chapter 3.6 on page 231.

Certain diseases can destroy the myelin sheath, causing a loss of coordination. Multiple sclerosis (MS) is a neurological condition, meaning it affects nerves. MS is caused by an autoimmune response to the myelin sheaths on nerve cells. The immune system starts to attack and destroy the myelin sheaths, which causes a loss of coordination and other neural functions as shown in **Figure 3.4.4**.

Figure 3.4.4 *Image summarising some of the words and ideas linked to the autoimmune disorder MS*

Glial cells

Glial cells

Cells that support neurons physically and chemically and produce myelin sheaths.

Glial cells are found in nervous tissue alongside neurons. These cells physically support neurons by holding them in a stable position and chemically support them by supplying nutrients to them. They also produce myelin sheaths as shown in **Figure 3.4.5**.

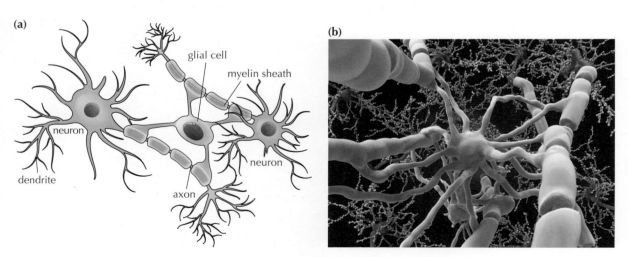

Figure 3.4.5 *Glial cells support neurons and are responsible for the myelination of long fibres (a) A glial cell in action (b) Visualisation of a glial cell in action*

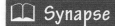

Neurotransmitters at synapses

It is perhaps surprising to learn that neurons in a neural pathway are separated from each other and from muscle fibre effectors by structures called **synapses**. Each synapse has a tiny gap called a synaptic cleft which can interrupt the passage of nervous impulses. Very rapid transmission across synapses is achieved using chemicals which temporarily fill the clefts.

> 📖 **Synapse**
>
> Region where a neuron joins to another neuron or to an effector cell.

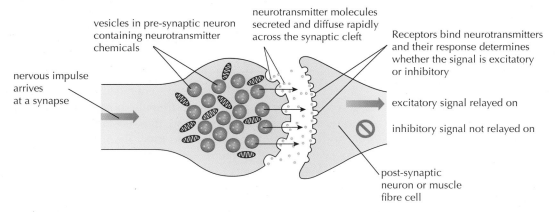

Figure 3.4.6 *A nervous impulse arrives at a synapse – note that the receptors determine if the signal is excitatory or inhibitory and relayed on or not*

In **Figure 3.4.6** a nervous impulse travels along an axon of a **pre-synaptic neuron** and arrives at a synapse. At the end of the axon there are vesicles in which chemicals called **neurotransmitters** are stored. The arrival of the impulse causes these vesicles to fuse with the pre-synaptic membrane of the neuron and secrete their contents into the synaptic cleft. The molecules of neurotransmitter diffuse rapidly across the cleft and bind to specific receptors in the **post-synaptic membrane** of the next neuron in the pathway. Receptors determine whether the signal is **excitatory** or **inhibitory** because binding the neurotransmitter molecules causes changes in the flow of electrical charges through the receptor. Some receptors change the flow to excite activity in the post-synaptic neuron and it is very likely that the impulse is relayed and flows on. Some, though, change the flow so that it is less likely that the impulse is relayed and the post-synaptic neuron in inhibited.

> 📖 **Neurotrans-mitters**
>
> Chemicals which are secreted into synapses to allow nervous impulses to pass.

> 🔍 **Hint**
>
> You must know that it is the receptors which determine whether the signal is excitatory or inhibitory but you don't have to know *how* they do this.

Deactivation of neurotransmitters

Neurotransmitters must be deactivated immediately their effect occurs so that the excitation or inhibition of the post-synaptic neuron is not permanent and the system remains sensitive and ready to respond to its next stimulation. Some neurotransmitters are broken down in the cleft by enzymes. For others, reuptake by the pre-synaptic membrane removes them from the cleft. This prevents continuous excitation or inhibition of post-synaptic neurons.

Transmission threshold

Synapses can effectively filter out weak stimuli arising from insufficient secretion of neurotransmitters. This is because a minimum number of weak stimuli from two different pre-synaptic neurons each cause the release of too few neurotransmitter molecules to reach threshold

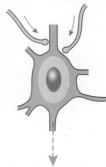

the stimuli are summated to reach the threshold and an impulse is generated in the post-synaptic neuron

Figure 3.4.7 *Summation of weak stimuli so that enough neurotransmitter is secreted to allow the impulse to be transmitted on*

neurotransmitter molecules must attach to receptors in order to reach the **transmission threshold** at which transmission of an impulse will occur. However, a series of weak stimuli from a number of pre-synaptic neurons can be **summated** to release enough neurotransmitter to trigger an impulse in the post-synaptic neuron. This occurs in convergent neural pathways where several neurons have synapses on to one neuron and can release enough neurotransmitter molecules to reach the threshold and trigger an impulse in it, as shown in **Figure 3.4.7**.

Neurotransmitter effects on mood and behaviour

Many different neurotransmitter chemicals have been discovered. They include the extremely powerful group of substances called **endorphins** and the substance **dopamine**. The study of neurotransmitters and their action has led to the discovery of drugs that alter neurotransmitter activity.

Endorphins

Endorphin production by the nervous system increases in response to severe injury, prolonged and continuous exercise, stress and certain foods. Endorphins stimulate the neurons involved in reducing the intensity of pain following injury. Increased levels of endorphins are also linked to the feelings of pleasure obtained from activities such as eating, sex and prolonged exercise.

The function of dopamine

The reward pathway in the brain is a group of neurons which is activated in response to dopamine. Dopamine is released when an individual engages in a behaviour that is beneficial to them, such as eating when hungry. The dopamine released activates the reward pathway which induces feelings of pleasure and reinforces the behaviour so that it potentially will be continued and repeated.

Agonist drugs

Many drugs used to treat neurotransmitter-related disorders are **agonists**. Agonists are drugs that bind to and stimulate specific receptors, mimicking the action of a neurotransmitter at a synapse as shown in **Figure 3.4.8**. Examples of agonists are heroin, methadone, morphine and opium.

Antagonist drugs

Antagonists are drugs that bind to specific receptors, blocking the action of a neurotransmitter at a synapse as shown in **Figure 3.4.8**. Examples of antagonists are the class of medications known as beta blockers. Beta blockers are predominantly used to manage abnormal heart rhythms and to protect the heart from a second heart attack after a first heart attack.

📖 **Transmission threshold**
Minimum number of neurotransmitter molecules secreted to allow nervous impulse to be transmitted on.

📖 **Summation**
The adding up of neurotransmitter molecules so that a series of weak stimuli can result in transmission of a nerve impulse.

📖 **Endorphins**
Neurotransmitters which stimulate neurons involved in reducing intensity of pain.

📖 **Dopamine**
A neurotransmitter which induces feelings of pleasure.

📖 **Agonist drug**
Drug which mimics the effect of a specific neurotransmitter.

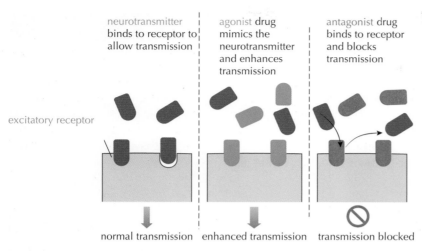

Figure 3.4.8 *Action of agonist and antagonist drugs at a synapse*

Other drugs

Other drugs act by inhibiting the enzymes that degrade neurotransmitters or by inhibiting reuptake of the neurotransmitter at the synapse, causing an enhanced effect. Acetylcholine esterase is an enzyme which breaks down the neurotransmitter acetylcholine in the synapse. Drugs which inhibit this enzyme can be used in the treatment of Alzheimer's disease. Drugs which inhibit the reuptake of noradrenaline from the synapse can be used in the treatment of depression.

Mode of action of recreational drugs

Recreational drugs affect neurotransmission at synapses in the brain, altering an individual's mood, cognition, perception and behaviour. Recreational drugs may be agonists or antagonists and many affect neurotransmission in the reward pathway of the brain. Many of these drugs are illegal in the UK.

Drug tolerance

Drug tolerance is caused by repeated use of drugs that act as agonists. Agonists stimulate specific receptors causing the nervous system to decrease both the number and sensitivity of these receptors. This **desensitisation** leads to **drug tolerance** where the individual must take more of the drug to get an effect.

Drug addiction

Drug addiction is caused by repeated use of drugs that act as antagonists. Antagonists block specific receptors causing the nervous system to increase both the number and sensitivity of these receptors. This is called **sensitisation** and can lead to **drug addiction** in which the individual craves more of the drug.

🔍 **Hint**

You don't need to learn examples of agonists but you must be able to say how they work.

📖 **Antagonist drug**

Drug which blocks the binding of a neurotransmitter to its receptors.

💥 **Make the link**

There is more about abnormal heart rhythms in Chapter 2.6 on page 157.

💥 **Make the link**

There is more about enzyme inhibition in Chapter 1.6 on page 65.

📖 **Drug tolerance**

When an individual requires more and more of a drug to gain the same effect as previously.

📖 **Drug addiction**

When an individual craves the drug to which they are addicted.

GO! Activity 3.4.1 Work individually to ...

Structured questions

1. **Give** the function of the following:
   a) An axon   1
   b) Myelination   1
2. **Describe two** methods of removal of a neurotransmitter from the synaptic cleft.   2
3. **Describe** the functions of the following neurotransmitters:
   a) Endorphins   1
   b) Dopamine   1
4. **Explain** what is meant by excitatory and inhibitory signals at a synapse.   3
5. **Explain** what is meant by 'summation' in the transmission of nervous impulses.   3
6. A trained and healthy male was asked to exercise by running on a treadmill for an extended period of time.

   The levels of endorphin and lactate in his bloodstream were measured over the period and are shown in the graph.

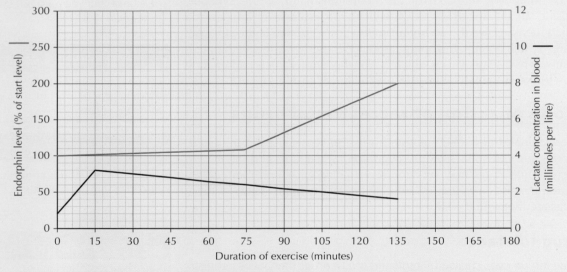

   a) **Use values from the graph to describe** how the endorphin level in the blood changes over the 135 minutes.   2
   b) **Predict** the endorphin level 165 minutes after the start of exercise.   1
   c) **Give** the lactate concentration in the blood when the endorphin level was 150% of its starting value.   1
   d) **Calculate** the percentage increase in lactate concentration from the start until 15 minutes into the exercise period.   1
   e) **Explain** how endorphins might help the individual to keep up exercise levels.   1
   f) **Explain** why the lactate levels in the blood are higher during exercise.   1

Make the link

There is more about lactate metabolism in Chapter 1.8 on page 83.

 Make the link

There is more about selecting and processing information and making predictions in Chapter 4 on page 263.

Extended response questions

1. **Give an account** of the structure and function of neurons.   5
2. **Give an account** of the transmission of a nervous impulse at a synapse.   7
3. **Write notes on** drugs and the nervous system under the following headings:
   a) Agonists   2
   b) Antagonists   2
   c) Drug tolerance and addiction   6

GO! Activity 3.4.2 Work in pairs to …

1. **Card sequencing activity: Events in neurotransmission**

   You will need: a stopwatch and a mini whiteboard with a marker pen.

   • Read the phrases in the grid below, which describe events in neurotransmission.

   • Work together to arrange the phrases in the correct order – start with phrase 2. Mark your answer on the whiteboard. Your teacher will check your work or put the correct sequence on the board.

   • You should each try this again individually against the clock – who is faster to the correct answer?

   • Your teacher may provide you with a photocopy of the grid. If so, cut it into strips, re-sequence it and glue it into your notes to make a permanent flowchart.

1 Neurotransmitter diffuses across synaptic cleft
2 Nerve impulse arrives at a pre-synaptic membrane
3 Vesicles release neurotransmitter into synaptic cleft
4 Vesicles move to and fuse with the pre-synaptic membrane
5 Neurotransmitter molecules bind to receptors on the post-synaptic membrane
6 Post-synaptic neuron either excited or inhibited
7 Neurotransmitter deactivated after action

2. **Flashcard activity**

   You will each need: a set of blank flashcards (A7 cards) and a stopwatch.

   • Find the glossary terms for this chapter – they are the **black** typeface and **red** typeface terms. Using your blank cards, you should each make a set of flashcards for these terms

(continued)

– write the term on one side and the definition on the other. You will find the definitions in the chapter.

- Shuffle your cards and lay them out in a column, some showing terms and some showing definitions – you decide. Your partner should match their cards with yours, laying their cards in a column beside yours to give the corresponding term or definition. Time how long they take to do this.
- Now swap roles – your partner should lay out their cards and you should try the matching exercise while your partner times you.
- You should each keep your set of flashcards as a revision tool for later.

(GO!) Activity 3.4.3 Work as a group to …

1. Research some different drugs.

Research the following drugs – you could focus on one each:

- Strychnine
- Curare
- Cannabis
- Cocaine

For each drug, find out about its mode of action and say if it is an agonist or an antagonist.

Share your findings and then each write an 80-word report on all four drugs.

2. Dice and Slice: Cells of the nervous system

You will need: Dice and Slice board (Appendix 2), the question and answer set and a dice.

- Take turns to play – have six turns each.
- Roll a dice for the top row number and again for the side number. Your partner will read you the question indicated by these numbers and you should try to answer it (your partner will tell you if you're right).
- If you get your question right, add your dice throw numbers to your score card and total them.
- After you have both had six turns you should find your own overall totals. Who has won?
- If you have time, play again and try to improve your scores.

Learning checklist

After working on this chapter, I can:

Knowledge and understanding

1. State that neurons are nerve cells.

2. State that neurons have a cell body and fibres called dendrites and axons.

3. State that three types of neuron are sensory, inter and motor.

4. State that axons are surrounded by a myelin sheath which insulates the axon and increases the speed of impulse conduction.

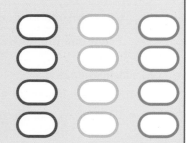

5. State that myelination continues from birth to adolescence.

6. State that responses to stimuli in the first two years of life are not as rapid or coordinated as those of an older child or adult.

7. State that certain diseases destroy the myelin sheath causing a loss of coordination.

8. State that glial cells produce the myelin sheath and physically support neurons.

9. State that synapses are gaps between neurons.

10. State that neurons connect with other neurons or muscle fibres at a synaptic cleft.

11. State that neurotransmitters relay impulses across the synaptic cleft.

12. State that neurotransmitters are stored in vesicles in the axon endings of the pre-synaptic neuron.

13. State that neurotransmitters are released into the synaptic cleft on arrival of an electrical impulse.

14. State that neurotransmitters diffuse across the synaptic cleft and bind to receptors on the membrane of the post-synaptic neuron.

15. State that neurotransmitters need to be removed by enzymes or reuptake to prevent continuous stimulation of post-synaptic neurons.

16. State that receptors determine whether the signal is excitatory or inhibitory.

17. State that synapses can filter out weak stimuli arising from insufficient secretion of neurotransmitters.

18. State that a minimum number of neurotransmitter molecules must attach to receptors in order to reach the threshold on the post-synaptic membrane to transmit the impulse.

19. State that summation of a series of weak stimuli can release enough neurotransmitter to trigger an impulse.

20. State that convergent neural pathways can release enough neurotransmitter molecules to reach threshold and trigger an impulse.

21. State that endorphins are neurotransmitters that stimulate neurons involved in reducing the intensity of pain.

22. State that endorphin production increases in response to severe injury, prolonged and continuous exercise, stress and certain foods.

23. State that increased levels of endorphins are also linked to the feelings of pleasure obtained from activities such as eating, sex and prolonged exercise.

24. State that dopamine is a neurotransmitter that induces feelings of pleasure and reinforces particular behaviours by activating the reward pathway in the brain.

25. State that the reward pathway involves neurons which secrete or respond to dopamine.

26. State that the reward pathway is activated when an individual engages in a behaviour that is beneficial to them, such as eating when hungry.

27. State that many drugs used to treat neurotransmitter-related disorders are agonists or antagonists.

28. State that agonists are chemicals that bind to and stimulate specific receptors, mimicking the action of a neurotransmitter at a synapse.

29. State that antagonists are chemicals that bind to specific receptors, blocking the action of a neurotransmitter at a synapse.

30. State that other drugs act by inhibiting the enzymes that degrade neurotransmitters or by inhibiting reuptake of the neurotransmitter at the synapse, causing an enhanced effect.

31. State that recreational drugs can also act as agonists or antagonists.

32. State that recreational drugs affect neurotransmission at synapses in the brain, altering an individual's mood, cognition, perception and behaviour.

33. State that many recreational drugs affect neurotransmission in the reward pathway of the brain.

34. State that drug addiction is caused by repeated use of drugs that act as antagonists.

35. State that antagonists block specific receptors, causing the nervous system to increase both the number and sensitivity of these receptors.

36. State that this sensitisation leads to addiction where the individual craves more of the drug.

37. State that drug tolerance is caused by repeated use of drugs that act as agonists.

38. State that agonists stimulate specific receptors, causing the nervous system to decrease both the number and sensitivity of these receptors.

39. State that this desensitisation leads to drug tolerance where the individual must take more of the drug to get an effect.

Skills

1. *Select information from a double axis line graph.*

2. *Process information from a double axis line graph.*

3. *Make a prediction by extrapolating from a double axis line graph.*

3.5 Non-specific body defences

You should already know:

- White blood cells are part of the immune system and are involved in destroying pathogens.
- Phagocytes carry out phagocytosis by engulfing and digesting pathogens.
- Pathogens are disease-causing microorganisms (bacteria, viruses, fungi).

Learning intentions

- Describe physical and chemical defences in terms of epithelial cells and chemical secretions produced against invading pathogens.
- Describe the inflammatory response and the role of mast cells, phagocytes and clotting elements.
- Describe the action of phagocytes and the role of cytokines as signalling molecules.

Introducing non-specific body defences

A **pathogen** is a microorganism that can cause disease, such as viruses, some bacteria and some fungi. The **non-specific immune system** is a generalised response to pathogenic microorganisms and includes physical barriers, chemical defences and the inflammatory response. These defences are not directed against any one pathogen but instead provide a guard against all infection. The strength of the non-specific general defence is that it is able to take action very quickly. Bacteria that have entered the skin through a small wound are detected and partly destroyed on the spot within a few hours. The limitation of the non-specific immune system is that it does not give the long-lasting or protective immunity of the specific immune system.

Physical barriers

The human body has several natural barriers to stop harmful microorganisms getting inside.

Closely packed **epithelial cells** form a physical barrier to pathogens. These are found in the skin and inner linings of the digestive and respiratory systems. **Figure 3.5.1** shows a layer of closely packed epithelial cells providing a physical barrier to the entry of pathogens.

> **Make the link**
>
> There is more about the specific immune response in Chapter 3.6 on page 231.

> **Non-specific immune system**
>
> A generalised response to pathogenic microorganisms that includes physical barriers, chemical defences and the inflammatory response.

> **Epithelial cells**
>
> A type of tissue made up of densely packed cells found in the skin and lining the respiratory and digestive systems.

epithelial cells —

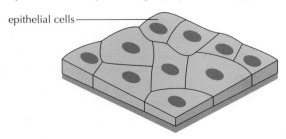

Figure 3.5.1 *Epithelium forms a physical barrier to the entry of pathogens*

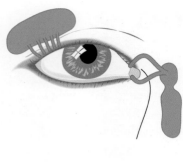

Figure 3.5.2 *The tear gland releases antiseptic tear fluid which washes over the eye and drains into the nose*

Chemical defence

Epithelial cells in cavity linings produce protective chemical secretions.

Chemical secretions are produced against invading pathogens. Secretions include tears, saliva, mucus and stomach acid. Tear glands under the eyelid produce tear fluid containing enzymes that kill bacteria that land on the surface of the eye. **Figure 3.5.2** shows the position of the tear gland under the eyelid and the tubes that drain the tear fluid into the back of the nose.

Saliva has antibacterial and antiviral properties. It contains proteins that bind to bacteria preventing them from attaching to any surfaces. Saliva also contains antimicrobial proteins which inhibit uncontrolled growth of bacteria.

Goblet cells in the trachea secrete mucus which traps microorganisms that we breathe in. Ciliated epithelium, as shown in **Figure 3.5.3**, sweep mucus and trapped microorganisms away from the lungs into the mouth, where they are then swallowed. The stomach produces a strong acid that kills most microorganisms present on our food or which enter our bodies when our hands touch our mouth.

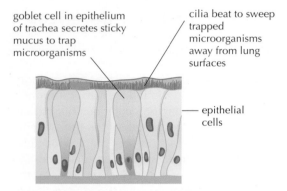

goblet cell in epithelium of trachea secretes sticky mucus to trap microorganisms

cilia beat to sweep trapped microorganisms away from lung surfaces

epithelial cells

Figure 3.5.3 *Ciliated epithelium of trachea, which provides a chemical barrier to microorganisms*

The inflammatory response

Pathogenic microorganisms can sometimes get past the first-line defence barriers. The **inflammatory response** (inflammation) occurs when tissues are injured by bacteria, trauma, toxins, heat or any other cause. **Figure 3.5.4** shows what happens during the inflammatory response following an injury. **Mast cells** are a type of white blood cell located at the boundaries between tissues and the external environment such as the skin. They play a key role in the inflammatory response. Following injury, mast cells release chemicals called **histamines** causing **vasodilation** of the surrounding blood vessels and increased capillary permeability. The increased blood flow leads to an accumulation of phagocytes and clotting elements at the site of infection. The phagocytes engulf and destroy the bacteria and the clotting elements promote wound healing. Phagocytes also release small protein molecules called **cytokines** which attract more phagocytes to the site of infection.

📖 **Inflammatory response**

A non-specific response to injury or infection involving vasodilation of surrounding blood vessels and increased capillary permeability.

📖 **Mast cells**

A type of white blood cell that releases histamine in response to tissue damage.

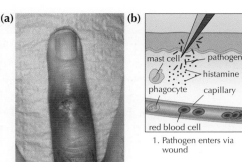

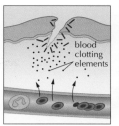

1. Pathogen enters via wound

2. Release of histamine by mast cells

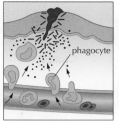

3. Histamine causes vasodilation and increased capillary permeability

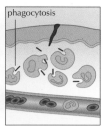

4. Increased blood flow and secretion of cytokines by phagocytes

5. Accumulation of phagocytes, specific white blood cells and clotting elements at site of infection

6. Wound clotted

7. Pathogen destroyed by phagocytosis

Figure 3.5.4 *(a) Red swelling at the site of an infection (b) The sequence of events (1-7) in the inflammatory response*

The action of phagocytes

Phagocytes recognise pathogens and destroy them by a process called **phagocytosis**.

Phagocytosis involves the engulfing of pathogens inside a food vacuole and their destruction by powerful digestive enzymes contained in vesicles called **lysosomes** as shown in **Figure 3.5.5**.

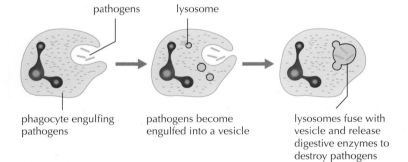

phagocyte engulfing pathogens

pathogens become engulfed into a vesicle

lysosomes fuse with vesicle and release digestive enzymes to destroy pathogens

Figure 3.5.5 *Stages in the process of phagocytosis*

Phagocytes release small protein molecules called cytokines that are important in cell signalling and communication. Cytokines act as a signal to recruit other phagocytes and some specific white blood cells, causing them to accumulate at the site of the infection.

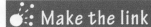

Make the link

There is more about blood clotting in Chapter 2.7 on page 166.

Hint

Remember that cytokines are cell signalling molecules and cause cells to accumulate at a site of infection.

📖 Histamines

Chemicals released by mast cells that cause vasodilation of the surrounding blood vessels and increased capillary permeability.

📖 Vasodilation

Increase in arteriole diameter to increase flow of blood.

📖 Cytokines

A group of signalling molecules released by a variety of cells including phagocytes that attract more phagocytes to the site of an infection.

📖 Phagocytosis

Process by which phagocytes engulf and destroy pathogens.

📖 Lysosomes

Vesicles or organelles in phagocytes containing digestive enzymes used to destroy pathogens.

GO! Activity 3.5.1 Work individually to ...

Structured questions

1. The non-specific immune system provides resistance to infection by physical, chemical and cellular means.

   a) **Name** the type of cell which forms a physical barrier to pathogens in the skin   1

   b) Mast cells initiate the inflammatory response.

      Name the chemical released by the mast cells and **describe** its role in the inflammatory response.   2

   c) **Describe** the role of cytokines in the inflammatory response.   1

   d) **Give two** examples of chemical secretions that provide defence against pathogens.   2

2. Leukocytosis is a condition indicated by an increased white blood cell count above the normal range in the blood, with specific cut-offs based on a patient's age. It is frequently a sign of an inflammatory response. One patient was diagnosed with leukocytosis when their white blood cell count increased temporarily to 11×10^9/litre. The graph shows the white blood cell count of the patient over a 20-week period.

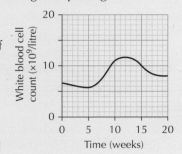

   a) **Calculate** the range in the white blood cell count over the 20-week period.   1

   b) **State** the number of weeks that the white blood cell count was equal to or greater than 10×10^9/litre.   1

   c) **Calculate** the average decrease in the white blood cell count between week 14 and week 20.   1

Extended response questions

1. **Give an account** of the role of phagocytes.   5
2. **Give an account** of non-specific body defences.   8

GO! Activity 3.5.2 Work in pairs to ...

1. **Card sequencing activity: The inflammatory response**

   You will need: a stopwatch and a mini whiteboard with a marker pen.

   - Read the phrases in the grid below, which describe the inflammatory response.

   - Work together to arrange the phrases in the correct order – start with phrase 7. Mark your answer on the whiteboard. Your teacher will check your work or put the correct sequence on the board.

   - You should each try this again individually against the clock – who is faster to the correct answer?

   - Your teacher may provide you with a photocopy of the grid. If so, cut it into strips, re-sequence it and glue it into your notes to make a permanent flowchart.

1 Mast cells release histamine
2 Accumulated phagocytes engulf and destroy bacteria

| 3 The increased blood flow and permeability leads to an accumulation of phagocytes and clotting elements at the site of infection |
| 4 Clotting elements promote wound healing |
| 5 Histamine causes surrounding blood vessels to vasodilate and increases capillary permeability |
| 6 Phagocytes release cytokines to attract more phagocytes to the site of the infection |
| 7 Tissues are injured by bacteria, trauma, toxins, heat or any other cause |

2. **Flashcard activity**

You will each need: a set of blank flashcards (A7 cards) and a stopwatch.

- Find the glossary terms for this chapter – they are the **black** typeface and red typeface terms. Using your blank cards, you should each make a set of flashcards for these terms – write the term on one side and the definition on the other. You will find the definitions in the chapter.

- Shuffle your cards and lay them out in a column, some showing terms and some showing definitions – you decide. Your partner should match their cards with yours, laying their cards in a column beside yours to give the corresponding term or definition. Time how long they take to do this.

- Now swap roles – your partner should lay out their cards and you should try the matching exercise while your partner times you.

- You should each keep your set of flashcards as a revision tool for later.

GO! Activity 3.5.3 Work as a group to …

1. **Placemat activity: Non-specific body defences**

You will need: a placemat template (Appendix 3) and four fine marker pens.

- Set the placemat in the middle of the table and each write your name in a section.

- Each participant should then write words that they think are related to *non-specific body defences* into their section of the placemat. Spend 2 minutes doing this.

- You should take it in turns to read out a word from your section. If everyone agrees it is related to the topic it can be copied into the centre section of the placemat. Continue until all words have been discussed.

- Working as a group, use all the words in the centre section to summarise your knowledge of non-specific body defences.

2. **Design and make an A2 flowchart collage to show the process of phagocytosis.**

You will need: an A2 sheet of paper, six pieces of different coloured card, scissors, a glue stick and a marker pen.

Work together to make a flowchart collage entitled 'The process of phagocytosis'. Make sure you label the parts of your flowchart.

Your teacher may ask your group to present your work to the class.

Learning checklist

After working on this chapter, I can:

Knowledge and understanding

1. State that a pathogen is a bacterium, virus or other organism that can cause disease.

2. State that non-specific defences can be physical and chemical.

3. State that epithelial cells form a physical barrier.

4. State that closely packed epithelial cells are found in the skin and inner linings of the digestive and respiratory systems.

5. State that chemical secretions such as tears, saliva, mucus and stomach acid are produced against invading pathogens.

6. State that the inflammatory response is a defence mechanism triggered by damage to living tissue.

7. State that histamine is released by mast cells causing vasodilation and increased capillary permeability. The increased blood flow leads to an accumulation of phagocytes and clotting elements at the site of infection.

8. State that phagocytes recognise pathogens and destroy them by phagocytosis.

9. State that phagocytosis involves the engulfing of pathogens and their destruction by digestive enzymes contained in lysosomes.

10. State that phagocytes release cytokines which attract more phagocytes to the site of infection.

11. State that cytokines are protein molecules that act as a signal to specific white blood cells, causing them to accumulate at the site of infection.

Skills

1. *Process information to calculate a range and average increase.*

2. *Select information from a graph.*

3.6 Specific cellular defences against pathogens

You should already know:

- Pathogens are disease-causing microorganisms (bacteria, viruses, fungi).
- Some lymphocytes produce antibodies which destroy pathogens.
- Each antibody is specific to a particular pathogen.

Learning intentions

- Describe the role and action of lymphocytes in the specific immune response.
- Describe the formation of clonal populations of identical lymphocytes.
- Describe the role of B-lymphocytes in terms of antibody production and an allergic reaction.
- Describe the role of T-lymphocytes in terms of apoptosis and autoimmune diseases.
- Explain the secondary response and the role of memory cells.
- Explain the effect of the human immunodeficiency virus (HIV) on the immune system.

The action of lymphocytes

Lymphocytes are the white blood cells involved in the specific immune response. Lymphocytes respond to specific antigens on invading pathogens. Antigens are molecules such as proteins, located on the surface of cells, which are able to trigger a **specific immune response**. It is the presence of antigens that are associated with specific pathogens that triggers the specific response to each individual pathogen. Each lymphocyte has a single type of membrane receptor that is specific to one antigen. Lymphocytes with membrane receptors that recognise an individual's own or self-antigens normally die during embryonic development. When a lymphocyte with membrane receptors that match an antigen on an invading pathogen arrives in the infected area, it binds with the antigen of the invading pathogen and is said to have been selected. The lymphocyte then divides repeatedly, resulting in the formation of a **clonal population** of identical lymphocytes. This is called **clonal selection**. Most of the clonal population are involved in fighting the infection but some remain in the body as **memory cells** as shown in **Figure 3.6.1**.

📖 Specific immune response

Immune system that responds to antigens by producing cells that directly attack the pathogen, or by producing special proteins called antibodies.

📖 Clonal population

Group of genetically identical cells produced from one parent cell.

📖 Clonal selection

Process in which B-lymphocytes or T-lymphocytes recognise a specific antigen and are then reproduced to produce a clonal population.

📖 Memory cell

A long-lived lymphocyte capable of recognising and responding to a particular antigen that has infected the body previously.

231

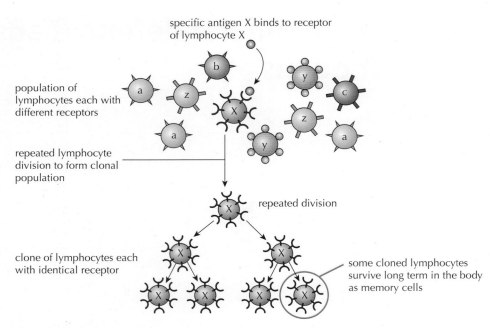

Figure 3.6.1 *The clonal selection of lymphocytes and formation of long-term memory cells*

There are two types of lymphocytes: B-lymphocytes and T-lymphocytes.

B-lymphocytes

B-lymphocytes produce **antibodies** against antigens and this leads to the destruction of the invading pathogen. Antibodies are Y-shaped proteins that have receptor binding sites specific to a particular antigen on a pathogen. Antibodies bind to the antigens and inactivate the pathogen. The resulting **antigen–antibody complex** can then be destroyed by phagocytosis as shown in **Figure 3.6.2**.

📖 **B-lymphocytes**

White blood cells with specific cell surface receptors that secrete specific antibodies into the blood and lymph.

📖 **Antigen– antibody complex**

The binding of an antibody to an antigen.

Make the link

There is more about phagocytosis in Chapter 3.5 on page 227.

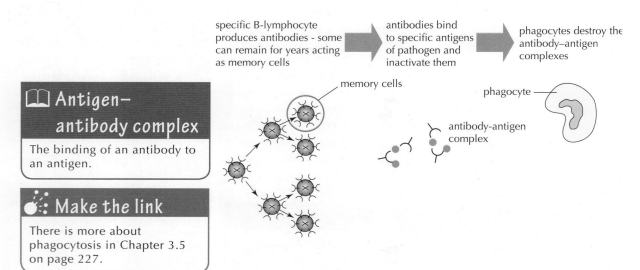

Figure 3.6.2 *The action of B-lymphocytes*

Allergy

B-lymphocytes can respond to antigens on substances that are harmless to the body, such as pollen. This **hypersensitive response** is called an **allergic reaction**.

T-lymphocytes

T-lymphocytes destroy infected body cells by recognising antigens of the pathogen on the cell membrane and inducing **apoptosis**. Apoptosis is programmed cell death. The T-lymphocytes attach on to the membrane of the infected cells and release proteins that diffuse into the cells. This results in the production of self-destructive enzymes which cause cell death. The remains of the cell are then removed by phagocytosis as shown in **Figure 3.6.4**.

Figure 3.6.3 *Hay fever allergy: a hypersensitive B-lymphocyte response to harmless pollen antigens*

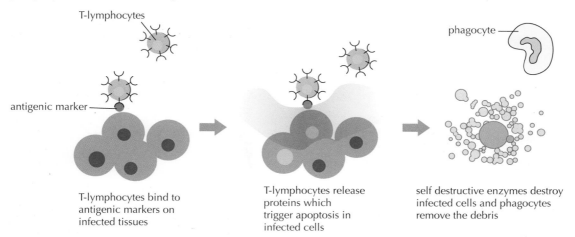

T-lymphocytes bind to antigenic markers on infected tissues

T-lymphocytes release proteins which trigger apoptosis in infected cells

self destructive enzymes destroy infected cells and phagocytes remove the debris

Figure 3.6.4 *The role of T-lymphocytes in apoptosis*

Autoimmune disease

T-lymphocytes can normally distinguish between **self-antigens** on the body's own cell membranes and **non-self-antigens** on the membranes of infected cells.

Failure of the regulation of the immune system leads to T-lymphocytes responding to self-antigens and is known as **autoimmunity**. In autoimmunity, the T-lymphocytes attack the body's own cells. This causes autoimmune diseases such as type 1 diabetes and rheumatoid arthritis.

In type 1 diabetes, the T-lymphocytes attack the insulin-producing cells in the pancreas, destroying them completely or damaging them enough to stop them producing insulin. Individuals with type 1 diabetes need daily injections of insulin.

In rheumatoid arthritis, the immune system mistakenly sends antibodies to the lining of the joints, where they attack the tissue surrounding the joint. This causes the thin layer of cells covering the joints to become sore and inflamed, releasing chemicals that damage nearby bones, cartilage, tendons and ligaments.

📖 Hypersensitive response

The response of B-lymphocytes to harmless substances such as pollen.

📖 Allergic reaction

The way the body reacts to the hypersensitive response of lymphocytes.

📖 T-lymphocytes

White blood cells that bring about the destruction of infected cells.

📖 Apoptosis

Process in which infected cells are induced to destroy themselves by T-lymphocytes.

233

Immunological memory

Some of the cloned B- and T-lymphocytes survive long term as memory cells. When a secondary exposure to the same antigen occurs, these memory cells rapidly give rise to a new clone of specific lymphocytes. These destroy the invading pathogens before the individual shows symptoms.

During the **secondary response**, antibody production is greater and more rapid than during the primary response, as shown in **Figure 3.6.5**.

The human immunodeficiency virus (HIV)

The human immunodeficiency virus (HIV) attacks and destroys T-lymphocytes as shown in **Figure 3.6.6**.

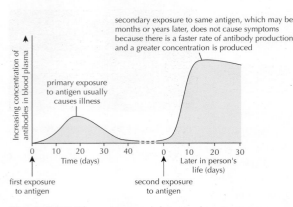

Figure 3.6.5 *The primary and secondary immune response*

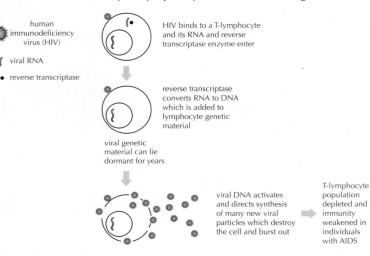

Figure 3.6.6 *Destruction and depletion of T-lymphocytes by HIV in individual with AIDS*

📖 Self-antigens

Antigens found on the surface of an individual's own body cells.

📖 Non-self-antigens

Antigens such as those found on the surface of pathogens which enter the body.

📖 Autoimmunity

A failure of regulation of the immune system that leads to T-lymphocytes responding to self-antigens and attacking the body's own cells.

Make the link

There is more about type 1 diabetes in Chapter 2.8 on page 177.

📖 Secondary response

When memory cells respond to a second or repeated exposure of an antigen which results in a faster production and greater concentration of antibodies.

HIV causes depletion of T-lymphocytes which leads to the development of **AIDS (acquired immune deficiency syndrome)**. Individuals with AIDS have a weakened immune system and so are more vulnerable to **opportunistic infections**.

📖 Opportunistic infections

Infection caused by pathogens that take advantage of an opportunity not normally available, such as a weakened immune system.

📖 AIDS (acquired immune deficiency syndrome)

A set of medical conditions which may result from a weakened immune system due to HIV infection.

🔍 Hint

Remember that HIV is a virus which enters the body and destroys T-lymphocytes so reducing the body's ability to defend itself against other infections.

🔵 Activity 3.6.1 Work individually to …

Structured questions

1. The diagram shows some of the stages leading to the production of a clone of B-lymphocytes by the immune system in response to infection by a pathogen.

   a) **Name** structures X.   1

   b) **Describe** what happens during Stage A.   1

   c) **Name** the chemicals which aid the movement of B-lymphocytes to the site of infection.   1

   d) The diagram shows how a clone of memory B-lymphocytes is produced.

   Describe one advantage of having memory cells.   1

   e) Failure of the immune system can lead to conditions such as allergies and autoimmune disease.

   i. **Give an account** of an allergic reaction.   2

   ii. **Give an account** of the autoimmune response.   2

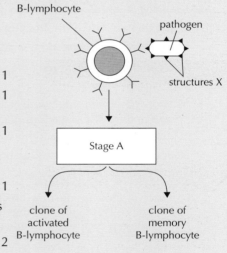

2. HIV is a virus which invades the cells of the immune system. People infected with HIV may not show symptoms for many years. AIDS is the condition which may develop from HIV infection, resulting in death.

   The graph shows the number of people in the world infected with HIV between 1990 and 2010 and the number of people who died from AIDS during the same period.

   a) **State** how many people were infected with HIV in the year 1995.   1

   b) **State** the number of deaths due to AIDS when 26 million people in the world were infected with HIV.   1

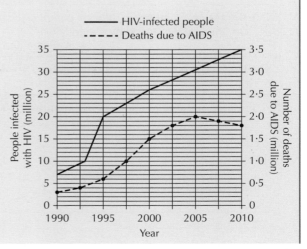

(continued)

c) **Calculate** the percentage increase in the number of people infected with HIV between 1995 and 2010. 1

d) **Calculate** the percentage of HIV-infected people who died from AIDS in 1995. 1

e) **Describe** the evidence from the graph which suggests that the treatment for individuals with HIV improved over this period. 1

> **Make the link**
>
> There is more about selecting and processing information from a double axis line graph in Chapter 4 on page 267.

Extended response questions

1. **Give an account** of the role of memory cells in the secondary immune response. 4
2. **Describe** the roles of T- and B-lymphocytes in specific immune response to disease. 8

Activity 3.6.2 Work in pairs to ...

1. **Card sort activity: Lymphocytes**

   You will need: a stopwatch and a mini whiteboard with a marker pen.

   - Read the phrases in the grid below, which describe features of B-lymphocytes and T-lymphocytes.
   - Work together to decide if statements 1–14 are features of B-lymphocytes or T-lymphocytes and make a list of each under the appropriate heading on the whiteboard. Your teacher will check your work or put the correct grouping on the board.
   - You should each try this again individually against the clock – who is faster to the correct answer?

1 These lymphocytes destroy infected body cells by recognising antigens of the pathogen on the cell membrane.	**8** Antibodies become bound to antigens, inactivating the pathogen.
2 Antibodies are specific to a particular antigen on a pathogen.	**9** The remains of the cell are then removed by phagocytosis.
3 The resulting antigen–antibody complex can then be destroyed by phagocytosis.	**10** These lymphocytes can respond to antigens on substances that are harmless to the body, such as pollen.
4 They induce programmed cell death.	**11** Released antibodies are Y-shaped proteins that have receptor binding sites.
5 Failure of the regulation of the immune system leads to these lymphocytes responding to self-antigens.	**12** These lymphocytes produce antibodies against antigens.
6 This causes autoimmune diseases such as type 1 diabetes and rheumatoid arthritis.	**13** They attach on to infected cells and release proteins that diffuse into the infected cells causing production of self-destructive enzymes which cause cell death.
7 This is called apoptosis.	**14** This hypersensitive response is called an allergic reaction.

2. Flashcard activity

You will each need: a set of blank flashcards (A7 cards) and a stopwatch.

- Find the glossary terms for this chapter – they are the **black** typeface and **red** typeface terms. Using your blank cards, you should each make a set of flashcards for these terms – write the term on one side and the definition on the other. You will find the definitions in the chapter.

- Shuffle your cards and lay them out in a column, some showing terms and some showing definitions – you decide. Your partner should match their cards with yours, laying their cards in a column beside yours to give the corresponding term or definition. Time how long they take to do this.

- Now swap roles – your partner should lay out their cards and you should try the matching exercise while your partner times you.

- You should each keep your set of flashcards as a revision tool for later.

GO! Activity 3.6.3 Work as a group to …

1. Placemat activity: Specific cellular defences

You will need: a placemat template (Appendix 3) and four fine marker pens.

- Set the placemat in the middle of the table and each write your name in a section.

- Each participant should then write words that they think are related to *specific cellular defences* into their section of the placemat. Spend 2 minutes doing this.

- You should take it in turns to read out a word from your section. If everyone agrees it is related to the topic it can be copied into the centre section of the placemat. Continue until all words have been discussed.

- Working as a group, use all the words in the centre section to summarise your knowledge of specific cellular defences.

2. Ring of Fire: Specific cellular defence

*You will need: a printout of the **Ring of Fire: Specific cellular defence** question and answer card set and a stopwatch.*

- Your teacher will deal the question and answer cards until all the cards have been issued.

- Your teacher will nominate a student to read the first question aloud and will start the clock.

- The student with the correct answer card should read the answer and then ask their own question.

- This is repeated until all the questions are completed. The timer is stopped.

- You should repeat the whole game to try to improve your time.

3. Research the causes, symptoms and treatment of hay fever and anaphylactic shock.

Visit the following web pages:

www.nhsinform.scot/illnesses-and-conditions/immune-system/hay-fever

www.nhsinform.scot/illnesses-and-conditions/immune-system/anaphylaxis

Split your group into two to work on the following topics:

- Causes and symptoms of hay fever
- Treatments for hay fever

(continued)

NEUROBIOLOGY AND IMMUNOLOGY

OR

- Causes and symptoms of anaphylactic shock
- Treatments for anaphylactic shock

Each group should produce a three-slide PowerPoint presentation on their topic and be prepared to present it to the class.

4. **Research the causes, symptoms and treatment of type 1 diabetes and rheumatoid arthritis.**

Visit the following web pages:

www.nhsinform.scot/illnesses-and-conditions/diabetes/type-1-diabetes

www.nhsinform.scot/illnesses-and-conditions/muscle-bone-and-joints/conditions/rheumatoid-arthritis

Split your group into two to work on the following topics:

- Causes and symptoms of type 1 diabetes
- Treatments for type 1 diabetes

OR

- Causes and symptoms of rheumatoid arthritis
- Treatments for rheumatoid arthritis

Each group should produce a three-slide PowerPoint presentation on their topic and be prepared to present it to the class.

5. **Research the control of HIV.**

Visit the following web pages:

www.nhsinform.scot/illnesses-and-conditions/immune-system/hiv#treating-hiv

www.hiv.scot/Pages/FAQs/Category/prep

Examine public health measures and drug therapies used in the control of HIV. Consider the topics below and discuss your findings as a group:

- Causes and symptoms of HIV
- Treatments for HIV
- Pre-Exposure Prophylaxis (or PrEP)

Learning checklist

After working on this chapter, I can:

Knowledge and understanding

1. State that lymphocytes are the white blood cells involved in the specific immune response.

2. State that lymphocytes respond to specific antigens on invading pathogens.

3. State that antigens are molecules, often proteins, located on the surface of cells that trigger a specific immune response.

4. State that lymphocytes have a single type of membrane receptor which is specific for one antigen. Antigen binding leads to repeated lymphocyte division resulting in the formation of a clonal population of identical lymphocytes.

5. State that there are two types of lymphocytes called B-lymphocytes and T-lymphocytes.

6. State that B-lymphocytes produce antibodies against antigens and this leads to the destruction of the pathogen.

7. State that antibodies are Y-shaped proteins that have receptor binding sites specific to a particular antigen on a pathogen.

8. State that antibodies become bound to antigens, inactivating the pathogen.

9. State that the resulting antigen–antibody complex can then be destroyed by phagocytosis.

10. State that B-lymphocytes can respond to antigens on substances that are harmless to the body, such as pollen. This hypersensitive response is called an allergic reaction.

11. State that T-lymphocytes destroy infected body cells by recognising antigens of the pathogen on the cell membrane and inducing apoptosis.

12. State that apoptosis is programmed cell death.

13. State that T-lymphocytes attach on to infected cells and release proteins.

14. State that these proteins diffuse into the infected cells causing the production of self-destructive enzymes which cause cell death. The remains of the cell are then removed by phagocytosis.

15. State that T-lymphocytes can normally distinguish between self-antigens on the body's own cells and non-self-antigens on infected cells.

16. State that failure of the regulation of the immune system leads to T-lymphocytes responding to self-antigens. This causes autoimmune diseases.

17. State that in autoimmunity, the T-lymphocytes attack the body's own cells.

18. State that this causes autoimmune diseases such as type 1 diabetes and rheumatoid arthritis.

Skills

1. *Process information to calculate a percentage and a percentage increase.*

2. *Select information from a graph.*

3.7 Immunisation

Learning intentions

- Explain how immunity can be developed by vaccination.
- Describe vaccines in terms of sources of antigens and the role of adjuvants.
- Explain the role of herd immunity in reducing the spread of diseases in terms of mass vaccination and the difficulty of achieving herd immunity thresholds.
- Explain the role and impact of antigenic variation in relation to the influenza virus.

Immunisation

📖 **Immunisation**

Process whereby an individual is made immune or resistant to an infectious disease.

Immunisation is the process whereby an individual is made immune or resistant to an infectious disease. Immunisation is a proven cost-effective health investment tool for controlling and eliminating life-threatening infectious diseases and is estimated to prevent between 2 and 3 million deaths each year.

Immunity can be developed by **vaccination**, using antigens from infectious pathogens to activate lymphocytes that produce memory cells.

 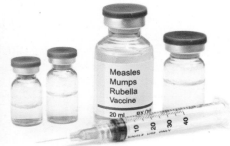

Figure 3.7.1 *The vaccine for measles, mumps and rubella*

📖 **Vaccination**

Using antigens from infectious pathogens to activate lymphocytes that produce memory cells.

📖 **Vaccine**

A substance used to stimulate the production of antibodies and provide immunity against one or several diseases.

The antigens used in **vaccines** can be inactivated pathogen toxins, dead pathogens, parts of pathogens and weakened pathogens, as shown in the following table.

Vaccine	Source of antigen
MMR (measles, mumps and rubella) Chickenpox	Weakened pathogen
HPV (human papillomavirus)	Parts of pathogen (individual proteins from the HPV)
Tetanus Diphtheria Whooping cough	Inactivated toxin from pathogen
Polio Rabies	Dead pathogen

Antigens are usually mixed with an **adjuvant** when producing the vaccine. An adjuvant is a substance that is mixed with the antigens in the vaccine to make it more effective. Aluminium salts are the most commonly used adjuvants in human vaccines. Aluminium adjuvants are used in the tetanus, HPV and hepatitis A and B vaccines but are not used in live viral vaccines. The adjuvant enhances or boosts the immune response to produce more antibodies and provide longer-lasting immunity. The use of an adjuvant can also reduce the amount of antigen needed in a vaccine and the number of doses that need to be given.

Herd immunity

Herd immunity describes how a population is protected from a disease after vaccination by stopping the pathogen responsible for the infection being transmitted between people, as shown in Figure 3.7.2. Transmission is interrupted by surrounding the infected person with vaccinated individuals. In this way even people who have not been vaccinated can be protected by reducing the probability that they will come into contact with infected individuals. Herd immunity occurs when a large percentage of a population is immunised. Establishing herd immunity is important in reducing the spread of diseases.

> **Make the link**
>
> There is more about memory cells in Chapter 3.6 on page 231.

> **Adjuvant**
>
> A substance that is mixed with the antigens in the vaccine to make it more effective and boost the immune system.

> **Herd immunity**
>
> Describes how a population is protected from a disease after vaccination by stopping the pathogen responsible for the infection being transmitted from person to person.

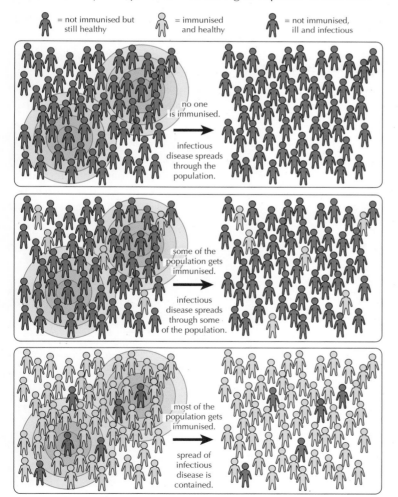

Figure 3.7.2 *Protection from spread of disease by herd immunity. Non-immune individuals are protected as there is a lower probability they will come into contact with infected individuals*

NEUROBIOLOGY AND IMMUNOLOGY

The percentage of immune individuals above which a disease may no longer circulate.

Hint

Remember there is no need to learn the actual herd immunity threshold for any of the diseases mentioned but you must know that that each disease does have a threshold.

The **herd immunity threshold** depends on the type of disease, the effectiveness of the vaccine and the density of the population. The following table shows the herd immunity threshold for different diseases.

Disease	Method of transmission	Herd immunity threshold (%)
Measles	Airborne	92–95
Whooping cough	Airborne droplet	92–94
Rubella	Airborne droplet	83–86
Polio	Faecal-oral route	80–86
Mumps	Airborne droplet	75–86

Mass vaccination programmes are designed to establish herd immunity to a disease and have been highly successful in reducing the 57 million deaths caused by infectious diseases in the world each year. Difficulties can arise when widespread vaccination is not possible due to poverty in the developing world, or when vaccines are rejected by a percentage of the population in the developed world. Adverse publicity regarding the safety of some vaccines can lead to a decrease in uptake and failure to reach the herd immunity threshold.

Antigenic variation

Some pathogens can change their antigens by a mechanism called **antigenic variation**. This means that memory cells are not effective against them.

Antigenic variation

Mechanism by which a pathogen alters its antigens to avoid a host's specific immune response via memory cells.

Antigenic variation occurs in the influenza (flu) virus. Creating a vaccine for a pathogen that can continually change its antigens is very difficult. This explains why influenza remains a major public health problem and why individuals who are at risk of infection require to be vaccinated every year.

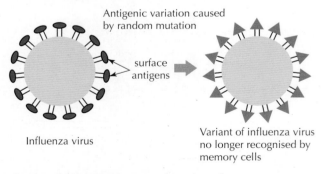

Figure 3.7.3 *Antigenic variation in the influenza virus*

🔵 GO! Activity 3.7.1 Work individually to ...

Structured questions

1. **a)** Safety concerns about the MMR vaccine caused the percentage of children in the UK immunised against measles, mumps and rubella to fall below the critical level of 80% between 2000 and 2005. As a result, outbreaks of these viral diseases occurred in various parts of the country.

   i. Give two sources of antigen that can be used in an injection of vaccine.   2

   ii. Explain how the process of vaccination prevents a child from showing symptoms of mumps during future outbreaks of the disease.   2

   iii. Suggest why these diseases spread more rapidly when the vaccination level falls below 80%.   1

   b) When producing a vaccine, an adjuvant is sometimes added.

   Explain the role of adjuvants in immunisation.   1

   c) Unlike the MMR vaccine, a vaccine against influenza needs to be given annually.

   State the reason for this.   1

2. The graph shows the number of cases of measles that occurred in the world between 1980 and 2010. It also shows the global vaccination rate against measles over the same period.

   a) State the vaccination rate when there were 3·1 million cases of measles in the world.   1

   b) State the increase in the vaccination rate for measles between 1980 and 1990.   1

   c) Calculate the average decrease in the number of cases of measles between 1980 and 1990.   1

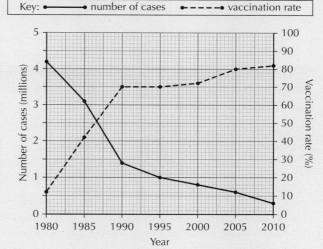

Key: ●——● number of cases   ●----● vaccination rate

   d) In many countries herd immunity has been established against measles.

   i. Describe what is meant by 'herd immunity'.   1

   ii. Suggest one reason why widespread vaccination programmes are not possible in all countries of the world.   1

   e) Predict when the number of cases of measles might be expected to reach zero if the trend from 2005–2010 was to continue.   1

3. In order to prevent transmission of measles occurring, it has been calculated that a herd immunity threshold of 93–95% is required. The table shows the percentage of the UK population aged 14 years and under who had received the measles vaccine by 1998 and 2003.

Year	Population under the age of 14 years (million)	Percentage vaccinated (%)
1998	11·2	91
2003	10·9	

(continued)

a) **Calculate** how many more children aged 14 and under would need to have been vaccinated in 1998 to have achieved the minimum herd immunity threshold. 　1

b) In 2003, in the same age group, 8·72 million children out of a total of 10·9 million were vaccinated against measles.

Calculate the percentage of children vaccinated in 2003. 　1

c) **Suggest** a reason for the decrease in the number of children vaccinated against measles in 2003. 　1

 **Make the link**

There is more about selecting, processing and predicting in Chapter 4 on page 263.

Extended response questions

1. **Give an account** of the principle of herd immunity. 　5
2. **Give an account** of immunisation under the following headings:
   a) Vaccination 　5
   b) Difficulties encountered in achieving widespread vaccination 　2

Activity 3.7.2 Work in pairs to ...

1. **Flashcard activity**

You will each need: a set of blank flashcards (A7 cards) and a stopwatch.

- Find the glossary terms for this chapter – they are the **black** typeface and **red** typeface terms. Using your blank cards, you should each make a set of flashcards for these terms – write the term on one side and the definition on the other. You will find the definitions in the chapter.

- Shuffle your cards and lay them out in a column, some showing terms and some showing definitions – you decide. Your partner should match their cards with yours, laying their cards in a column beside yours to give the corresponding term or definition. Time how long they take to do this.

- Now swap roles – your partner should lay out their cards and you should try the matching exercise while your partner times you.

- You should each keep your set of flashcards as a revision tool for later.

Activity 3.7.3 Work as a group to ...

1. **Placemat activity: Immunisation**

You will need: a placemat template (Appendix 3) and four fine marker pens.

- Set the placemat in the middle of the table and each write your name in a section.

- Each participant should then write words that they think are related to *immunisation, mass vaccination and antigenic variation* into their section of the placemat. Spend 2 minutes doing this.

- You should take it in turns to read out a word from your section. If everyone agrees it is related to the topic it can be copied into the centre section of the placemat. Continue until all words have been discussed.

- Working as a group, use all the words in the centre section to summarise your knowledge of immunisation, mass vaccination and antigenic variation.

2. Research vaccinations.

Visit the following website:

www.nhs.uk

Your task is to investigate some vaccinations given to people in the UK by the NHS.

On the home page of the website, select the tab 'Health A–Z', then select 'V' from the alphabetical list. Select 'Vaccinations' and then select 'NHS vaccinations and when to have them'.

Choose to research either the 6-in-1 vaccine given to babies under 1 year old or the HPV vaccine. Discuss the material on the web page and note down the answers to the questions below.

6-in-1 vaccine given to babies under 1 year old

a) **Name** the condition(s) the vaccine protects the baby from.

b) **Describe** how the vaccine is given.

c) **Explain** why the baby is given three doses.

d) **Describe** how well the vaccine works.

e) **Discuss** the safety of the vaccine.

HPV vaccine

a) **Name** the disease that this vaccine protects against.

b) **State** what HPV is.

c) **State** the age at which girls and boys are offered this vaccination.

d) **Describe** how the vaccine is given.

e) **State** how long the HPV vaccine protects for.

3. Research travel vaccinations.

Visit the following website:

www.nhs.uk

Your task is to investigate some travel vaccinations given to people in the UK by the NHS.

On the home page of the website, select the tab 'Health A–Z', then select 'T' from the alphabetical list. Select 'Travel vaccinations' and follow the link 'NHS Fit for Travel' under 'Which travel vaccines do I need?'

Answer the questions below:

a) Select a country and then find out which vaccinations are needed or recommended for that area.

b) From the 'Travel vaccinations' page, **name** the country that needs proof that visitors arriving for the Hajj and Umrah pilgrimages have been vaccinated against meningitis.

c) **Give two** examples of travel vaccinations that are usually free on the NHS.

d) If you need a travel vaccine, you are asked to see your doctor at least eight weeks before you travel. **Give one** reason for this.

Learning checklist

After working on this chapter, I can:

Knowledge and understanding

1. State that immunity can be developed by vaccination using antigens from infectious pathogens, so creating memory cells.

2. State that the antigens used in vaccines can be inactivated pathogen toxins, dead pathogens, parts of pathogens and weakened pathogens.

3. State that antigens are usually mixed with an adjuvant when producing the vaccine.

4. State that an adjuvant is a substance which makes the vaccine more effective, so enhancing the immune response.

5. State that herd immunity occurs when a large percentage of a population is immunised.

6. State that establishing herd immunity is important in reducing the spread of diseases.

7. State that non-immune individuals are protected as there is a lower probability they will come into contact with infected individuals.

8. State that the herd immunity threshold depends on the type of disease, the effectiveness of the vaccine and the density of the population.

9. State that mass vaccination programmes are designed to establish herd immunity to a disease.

10. State that difficulties can arise when widespread vaccination is not possible due to poverty in the developing world, or when vaccines are rejected by a percentage of the population in the developed world.

11. State that some pathogens can change their antigens. This means that memory cells are not effective against them.

12. State that antigenic variation occurs in the influenza virus. This explains why it remains a major public health problem and why individuals who are at risk require to be vaccinated every year.

Skills

1. *Select information from a double axis line graph.*

2. *Process information to calculate an average increase and a percentage.*

3. *Predict from a double axis line graph.*

3.8 Clinical trials of vaccines and drugs

Learning intentions
- Describe the role and design of clinical trials in establishing the safety and effectiveness of vaccines and drugs.

Clinical trials

Clinical trials are carried out to evaluate the effectiveness and safety of medications such as vaccines and drugs by monitoring their effects on large groups of people before they are approved and given a license for use.

Clinical trials may be conducted by government health agencies, researchers affiliated with a hospital or university medical program, independent researchers or private industry.

The design of clinical trials to test vaccines and drugs involves **randomised**, **placebo-controlled** and **double-blind protocols**.

Randomised protocols

Participants in clinical trials are usually volunteers, though in some cases research subjects may receive payment for their participation. Subjects are generally divided into two or more groups, including a **test group** and a **control group** that does not receive the experimental treatment and receives a **placebo** (inactive substance) instead. Subjects in clinical trials are divided into groups in a randomised way. This is done to reduce any bias in the distribution of characteristics in the groups, such as age and the gender of the participants.

Placebo-controlled protocols

One group of subjects receives the vaccine or drug while the second group receives a placebo control. This is done to ensure that valid comparisons can be made when evaluating the results.

Double-blind protocols

In a double-blind trial neither the subjects nor the researchers know which group the subjects are in. The subjects and the researchers are unaware which subjects are receiving the new treatment and which are being given the placebo. This is done to prevent a biased interpretation of the results by the researcher.

📖 Clinical trial
A research study carried out to obtain data about new drugs, vaccines or other treatments.

📖 Randomised protocol
Methods used to reduce bias in clinical trials by eliminating the effects caused by variables, such as age or gender of the participants.

📖 Placebo-controlled protocol
Types of control used in a clinical trial in which a random group of participants is given an inactive substance or blank rather than the treatment under trial.

📖 Double-blind protocol
Clinical trials in which neither the participants nor the researchers know which participants are given the treatment and which are the control group.

Make the link
There is more about vaccines in Chapter 3.7 on page 240.

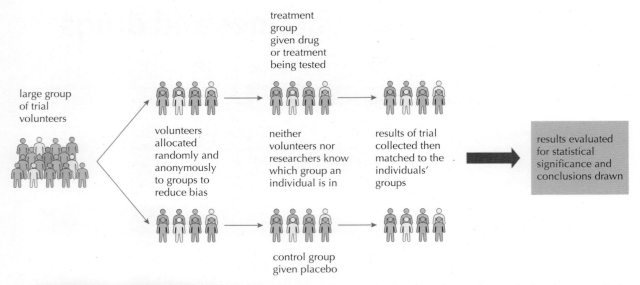

treatment group given drug or treatment being tested

large group of trial volunteers

volunteers allocated randomly and anonymously to groups to reduce bias

neither volunteers nor researchers know which group an individual is in

results of trial collected then matched to the individuals' groups

results evaluated for statistical significance and conclusions drawn

control group given placebo

Figure 3.8.1 *Summarises the features of a randomised, placebo-controlled, double-blind clinical trial*

📖 Statistical significance

The likelihood that a relationship between two or more variables is caused by something other than chance.

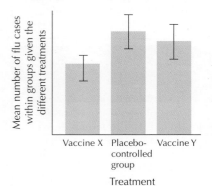

Figure 3.8.2 *Statistically processed results of a clinical trial of two new vaccines X and Y compared with a placebo control*

🔍 Hint

Remember that having a large sample size reduces experimental error and allows statistical tests to be carried out to determine the significance of the findings.

Group size

It is important that the group size in any clinical trial is large in order to increase the reliability of the results. A large group size can help to reduce experimental error and increases the **statistical significance** of the results. A larger sample size allows the researchers to increase the significance level of the findings, since the confidence of the result is likely to increase with a higher sample size.

At the end of the clinical trial, results from the two groups are compared to determine whether there are any statistically significant differences between the groups. The groups must be of a suitable size to reduce the magnitude of experimental error.

Statistical significance

Results of clinical trials can be processed statistically and then analysed to show if there are any differences between trial groups and placebo-controlled groups. **Figure 3.8.2** shows mean results from a clinical trial of two new influenza vaccines and the error bars shown in the chart give a statistical indication of the range of data obtained. Compared with the placebo-controlled group, the results for Vaccine X are statistically significant because the error bars for the two groups do not overlap. However, the data for Vaccine Y compared with the placebo-controlled group show that the results are not statistically significant because the error bars overlap.

While important and highly effective in preventing obviously harmful treatments from coming to market, clinical research trials are not always perfect. Side effects associated with long-term use and interactions between the experimental drug or vaccine and other medications may not be discovered during a trial.

 GO! Activity 3.8.1 Work individually to …

Structured questions

1. Medical research studies involving people are called clinical trials.

   a) **Explain** why new vaccines and drugs are subjected to clinical trials before being approved and given a license. 2

   b) **Explain** why it is good practice to create randomised groups in a clinical trial. 2

   c) Three thousand volunteers took part in a clinical trial to investigate the effect of a cholesterol-reducing drug. Half of the participants were given the cholesterol-reducing drug while half were given a placebo.

      i. **Describe** how a double-blind design could be achieved when setting up the clinical trial. 1

      ii. **State** what aspect of the design of the study increased the reliability of the results. 1

2. A drug to relieve the effects of severe migraine headaches was given to a group of 40 volunteers participating in a randomised, placebo-controlled clinical trial. The participants self-administered the medication as instructed at the onset of their headache and recorded the pain intensity experienced, with a score of between 0 and 10, every hour over a period of 4 hours.

   The results are shown in the table. $8\cdot0 \pm 1\cdot6$ means that the average of the data is $8\cdot0$ but that the data ranges between $6\cdot4$ and $9\cdot6$.

Time (hours)	Average pain intensity	
	Treatment group	Placebo group
0	$8\cdot0 \pm 1\cdot6$	$7\cdot8 \pm 1\cdot7$
1	$7\cdot2 \pm 1\cdot6$	$7\cdot7 \pm 1\cdot6$
2	$6\cdot0 \pm 1\cdot5$	$7\cdot2 \pm 1\cdot4$
3	$4\cdot0 \pm 1\cdot4$	$6\cdot6 \pm 1\cdot4$

   a) **Give two** factors which should be taken into account when constructing the groups to minimise any selection bias. 2

   b) **Calculate** the range in average pain intensity between the treatment group and the placebo group at 0 hours. 1

   c) **Construct a bar graph,** including error bars and a key, to show the ranges of data obtained and the results of this clinical trial. 3

   d) From the data, it was concluded that the treatment had proved effective.

      Describe evidence from the data that supports this conclusion. 1

   e) **Suggest two** reasons why the results of the trial may not be reliable. 2

   > ### ⚬⁚ Make the link
   > There is more about selecting, presenting, processing, planning and evaluating the reliability of a procedure in Chapter 4 on page 263.

Extended response question

1. **Describe** the role of design protocols for a clinical trial of a new vaccine or drug. 8

GO! Activity 3.8.2 Work in pairs to …

1. **Flashcard activity**

   You will each need: a set of blank flashcards (A7 cards) and a stopwatch.

   - Find the glossary terms for this chapter – they are the **black** typeface and **red** typeface terms. Using your blank cards, you should each make a set of flashcards for these terms – write the term on one side and the definition on the other. You will find the definitions in the chapter.

   - Shuffle your cards and lay them out in a column, some showing terms and some showing definitions – you decide. Your partner should match their cards with yours, laying their cards in a column beside yours to give the corresponding term or definition. Time how long they take to do this.

   - Now swap roles – your partner should lay out their cards and you should try the matching exercise while your partner times you.

   - You should each keep your set of flashcards as a revision tool for later.

GO! Activity 3.8.3 Work as a group to …

1. **Placemat activity: Clinical trials**

   You will need: a placemat template (Appendix 3) and four fine marker pens.

   - Set the placemat in the middle of the table and each write your name in a section.

   - Each participant should then write words that they think are related to *clinical trials* into their section of the placemat. Spend 2 minutes doing this.

   - You should take it in turns to read out a word from your section. If everyone agrees it is related to the topic it can be copied into the centre section of the placemat. Continue until all words have been discussed.

   - Working as a group, use all the words in the centre section to summarise your knowledge of clinical trials.

Learning checklist

After working on this chapter, I can:

Knowledge and understanding

1. State that vaccines and drugs are subjected to clinical trials to establish their safety and effectiveness before being licensed for use. ◯ ◯ ◯

2. State that the design of clinical trials to test vaccines and drugs involves randomised, double-blind and placebo-controlled protocols. ◯ ◯ ◯

3. State that subjects in clinical trials are divided into groups in a randomised way to reduce bias in the distribution of characteristics such as age and gender. ◯ ◯ ◯

4. State that in a double-blind trial neither the subjects nor the researchers know which group subjects are in to prevent bias.

5. State that one group of subjects in the trial receives the vaccine or drug while the second group receives a placebo control to ensure valid comparisons.

6. State that at the end of the trial, results from the two groups, which must be of a suitable size to reduce the magnitude of experimental error, are compared to determine whether there are any statistically significant differences between the groups.

Skills

1. *Process information to calculate a range.*

2. *Select information from a table and a graph.*

3. *Present data as a bar graph including error bars.*

4. *Identify variables in a procedure.*

5. *Suggest ways of improving reliability.*

Chapter 3 practice area test Neurobiology and immunology

Write your answers on separate sheets of paper. Mark your work using the answers online at www.collins.co.uk/pages/Scottish-curriculum-free-resources.

Paper 1: Multiple choice

Total: 10 marks

1. Which row in the table identifies components of the central and peripheral nervous systems?

	Central nervous system (CNS)		Peripheral nervous system (PNS)	
A	sympathetic	parasympathetic	brain	spinal cord
B	brain	somatic	spinal cord	autonomic
C	brain	spinal cord	autonomic	somatic
D	somatic	sympathetic	parasympathetic	autonomic

2. The diagram below represents a section through the brain.

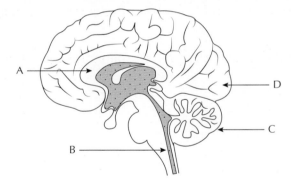

   Which letter indicates the cerebral cortex?

3. Some individuals suffering head injuries forget events that happened a few seconds before the injury occurred.

   This memory loss is most likely to be due to the injury affecting

   A retrieval

   B displacement

   C long-term memory

   D short-term memory.

4. The following statements refer to events that may occur at a synapse.

   1. An impulse is generated

   2. Neurotransmitter diffuses across the synaptic cleft

   3. Neurotransmitter is broken down by an enzyme

   4. Neurotransmitter is released from storage vesicles

   In which sequence do these events occur?

   A 1, 2, 4, 3

   B 2, 4, 3, 1

   C 4, 2, 1, 3

   D 4, 2, 3, 1

5. Which statement about the action of recreational drugs on neurotransmission is correct?

   A Sensitisation results from an increase in the number of neurotransmitter receptors due to the use of drugs that are agonists.

   B Sensitisation results from an increase in the number of neurotransmitter receptors due to the use of drugs that are antagonists.

   C Desensitisation results from an increase in the number of neurotransmitter receptors due to the use of drugs that are agonists.

   D Desensitisation results from an increase in the number of neurotransmitter receptors due to the use of drugs that are antagonists.

6. Which of the following is **not** part of the inflammatory response?

   A Production of antibodies

   B Increased capillary permeability

   C Vasodilation

   D Release of histamine

7. Failure in regulation of the immune system leading to an autoimmune disease is caused by a

   A T-lymphocyte immune response to self-antigens

   B B-lymphocyte immune response to self-antigens

   C T-lymphocyte immune response to foreign antigens

   D B-lymphocyte immune response to foreign antigens.

8. On which of the following does the herd immunity threshold not depend?

    A Type of disease

    B Effectiveness of the vaccine

    C Population density

    D Isolation of non-immune individual

9. The graph below shows the number of cases of meningitis and deaths due to meningitis in the UK from 1998 to 2001.

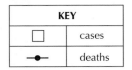

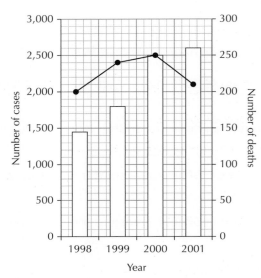

Year

    In which years were the number of deaths from meningitis greater than 10% of the number of cases?

    A 1998 and 1999

    B 1999 and 2000

    C 2000 and 2001

    D 1998, 1999 and 2001

10. Two groups of volunteers were used when carrying out clinical trials of a drug. One group was given the drug while the other group was given a placebo.

    The purpose of the placebo was to

    A reduce experimental error

    B ensure a valid comparison

    C allow a statistical analysis of the results to be carried out

    D ensure that researchers are unaware who has been treated with the drug.

Paper 2: Structured and extended response

Total: 40 marks

1. The image shows a vertical section through a human brain.

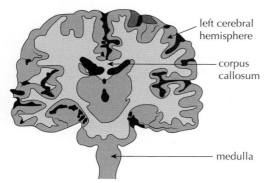

a) **State** the function of the motor area in the left cerebral hemisphere.　**1**

b) **State** the function of the corpus callosum.　**1**

c) i. **Name** the division of the nervous system with centres in the medulla.　**1**

ii. **Give one** function of the parasympathetic nervous system.　**1**

2. The diagram shows parts of some cells in the central nervous system (CNS).

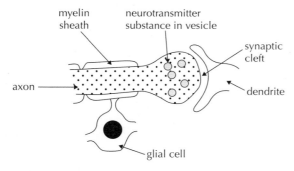

a) i. **Describe** the function of the myelin sheath.　**1**

ii. **Give one** function of glial cells.　**1**

b) i. **Describe** how neurotransmitters are involved in the passage of nervous impulses across the synaptic cleft.　**1**

ii. **Name** the neurotransmitters involved in the reduction of pain intensity following a trauma.　**1**

3. The diagram shows two different neural pathways. The arrows indicate the direction of the nerve impulses.

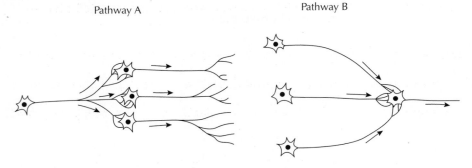

Pathway A   Pathway B

a) **Name** the types of pathway represented by A and B.   1

b) Pathways like A are involved in the complex function of the human hand.

   Explain how pathways like A control the function of the hand.   1

c) Neurotransmitters are secreted into synaptic clefts to allow nervous impulses to cross.

   Give one way in which the neurotransmitters are removed to prevent continuous stimulation of post-synaptic neurons.   1

4. The graph shows the percentages of children given the MMR vaccine between 2000 and 2010 in a health board area in the UK.

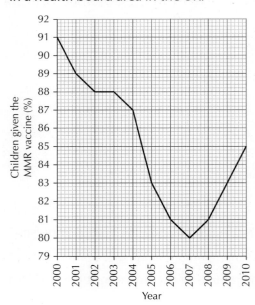

a) **Use values from the graph to describe** the changes in percentage MMR vaccination between 2004 and 2010. **1**

b) **Calculate** the average decrease per year in the percentage vaccination rate between 2000 and 2007. **1**

c) **Identify** the **two** years between which the unvaccinated percentage was less than 19%. **1**

d) Herd immunity threshold for this disease in this health board area is 87% vaccinated.

   Calculate the number of years during which the population was not protected by herd immunity. **1**

5. The diagram shows links between some processes involved in memory.

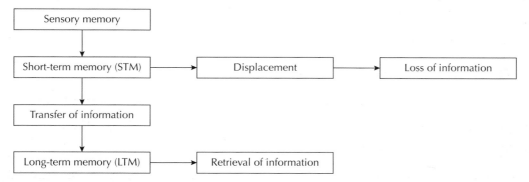

a) **Explain** why some information can be displaced from the STM and lost as shown in the diagram. **1**

b) **Name one** method for transferring information from the STM to the LTM. **1**

c) Retrieval of information can be aided by the use of contextual cues.

   Explain what is meant by a 'contextual cue'. **1**

6. An investigation was carried out on the effect of loud music on working memory by testing the ability of students to perform calculations. Twenty students were divided into two equal groups, A and B. Each group was given 20 calculations to complete.

   Group A sat in an evenly lit, quiet room.

   Group B sat in an evenly lit room where loud music was being played continuously.

   The average numbers of errors the students made while doing the calculations are shown in the table.

Student	Average number of errors	
	Group A	Group B
1	3	8
2	2	7
3	1	3
4	2	8
5	2	5
6	4	6
7	1	8
8	0	4
9	2	6
10	3	5

a) **Give** a hypothesis that could be tested by this experiment.   **1**

b) i. **Identify** the dependent variable in this investigation.   **1**

   ii. **Identify one** variable, not already indicated, which would have to be kept constant to ensure a valid comparison between the two groups could be made.   **1**

c) **Calculate** by how many times the average number of errors increased as a result of the loud music.   **1**

d) **Give** the meaning of the term 'working memory'.   **1**

e) A third group of 10 students carried out the investigation under the same conditions as group B, but were tested six times instead of only once. Each test comprised different calculations. The average percentage of errors is shown in the table.

Test	1	2	3	4	5	6
Average percentage of errors	36	31	24	22	21	21

   Construct a line graph to show the data from the table for the third group.   **2**

7. The diagram shows skin cells which have been pierced by a sharp rose thorn and a nearby blood capillary.

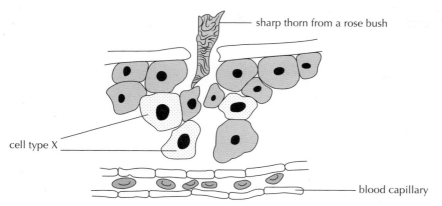

a) **Identify** cell type X which releases histamines following damage to the tissues.   **1**

b) **Describe one** effect of histamine.   **1**

c) Phagocytes and specific white blood cells accumulate at the site of infection.

   Name one other substance that accumulates to aid healing.   **1**

8. The diagram shows the structure of one strain of the influenza virus.

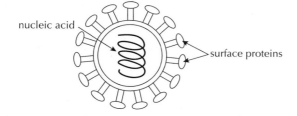

a) This virus can be used to prepare an influenza vaccine.

To do this the viral nucleic acid must be broken up to prevent the virus replicating but viral surface proteins must be left intact.

Explain why it is necessary to leave the surface proteins intact.   **1**

b) **Explain** the role of adjuvants in immunisation.   **1**

c) A different vaccine is required against each different strain of the influenza virus.

Explain why different vaccines are required.   **1**

d) Clinical trials of vaccines use a randomised protocol to avoid bias.

Describe how randomisation is achieved.   **1**

9. **Describe** how recreational drugs can affect the brain.   **8**

CHAPTER 4
Skills of scientific inquiry

Skills of scientific inquiry

You should already know:

Skills of scientific inquiry for Higher Human Biology are grouped under more or less the same headings as at National 5 Biology and you should already be familiar with the following categories:

- Planning and designing experiments to test hypotheses.
- Selecting information from a variety of sources.
- Presenting information appropriately in a variety of forms.
- Processing information using calculations and units, where appropriate.
- Making predictions and generalisations based on evidence or supplied information.
- Drawing valid conclusions and giving explanations supported by evidence and justifications.
- Evaluating and suggesting improvements to experimental methods.

Learning intentions

- To increase the levels of the skills listed above to take them to the level required for Higher Human Biology.

Make the link

There is more about the assignment in Chapter 5 on page 280.

Introducing skills of scientific inquiry

Scientific inquiry skills are a major part of the Higher Human Biology course. About a quarter of all the marks available for the exam question paper focus on skills as well as about three-quarters of the marks for your assignment.

Although the skills marks are scattered through the question paper, there will always be a large experimental design question and a large data-handling question. Experimental design and data-handling overlap a little so certain skills may sometimes be tested in either setting.

There are specific pieces of apparatus and specific experimental techniques with which you should be familiar and which could be asked about in exam questions and could feature in your assignment. These are detailed below.

Questions about the application of skills are very varied – it's impossible to give examples of every possible question. The information and examples below cover all of the skills but cannot cover every variation – there will always be something a bit different for you to think about and apply your skills to.

You should note that there are also examples of skills questions in each of the chapters of this book – these are always the last question or two in the 'Work individually' part of the Activities section for each chapter.

Experimental design skill set

This is a skill set connected to the design and planning of experimental procedures which are part of the scientific method of investigating problems.

(a)

(b)

Figure 4.1 *(a) The scientific method is an agreed set of principles used in carrying out investigative work (b) It involves carefully controlled experiments which are done in laboratories as part of research or in clinical trials*

In your exam there will be a large experimental design question worth between 5 and 9 marks which is likely to focus on an experiment which has been carried out. Experiments are designed to test a hypothesis. A hypothesis predicts a relationship between variables and all experiments should have an aim which links variables. The question should concentrate on **planning**, particularly **aims**, **variables**, **control**, **reliability** and **evaluation**. There may be a graph to **present** and some other skills may be tested.

Devising an aim

The aim of an experiment is usually given in the stem of an experimental question. If it is not given, you could be asked to state it. An aim should mention the independent and dependent variables.

Identifying variables

Variables are experimental factors which can change or be changed. Several potential variables are usually mentioned or suggested in the stem of a question. The stem can be complex.

Example

An experiment was carried out into the effects of exercise level on the rate of respiration. Five volunteers each undertook different exercise levels: sitting, walking, jogging, running and sprinting. After the activity, each volunteer breathed out into a bottle containing bicarbonate indicator which gradually turns yellow in the presence of carbon dioxide, as shown in the diagram. The time taken for the indicator to change to yellow depends on the level of carbon dioxide which in turn depends on the rate of respiration. The faster the indicator

Make the link

There is more about clinical trials in Chapter 3.8 on page 247.

Aim

Statement of the purpose of an experiment; it should include mention of the independent and dependent variables.

Variable

A factor which can change or be changed in an experimental setting.

Reliability

The level of confidence or trust in results which should ensure that it is the independent variable that has caused changes to the dependent variable and not chance or a fluke.

changes to yellow, the more carbon dioxide is present so the faster the rate of respiration.

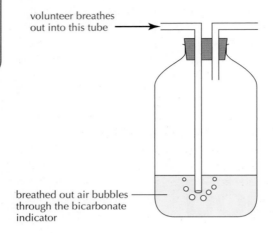

volunteer breathes out into this tube →

breathed out air bubbles through the bicarbonate indicator

Controlled variables are those which are kept the same throughout. These are the variables which are not included in the results table. Usually, some important variables are not mentioned in the description of the experiment and you could be asked to identify one or two of these missing ones. Validity in an experiment is enhanced by the control of variables. Which variable should be controlled in this example? Think about the volunteers – they should undertake the activities for the same length of time, at the same air temperature and wearing the same kit. They should breathe into the tube at the same time after activity and at the same rate. You may be able to think of other variables.

The **independent variable** is the variable being investigated in the aim – it should be in the first column of the results data table. You could be asked to identify it. In this example, the independent variable is the exercise level as mentioned in the aim.

The **dependent variable** is the variable which varies as an experiment result. It should be in the last column of the results data table. You could be asked to identify it. In this example, the dependent variable is the time taken for the indicator to turn yellow, as mentioned in the question stem.

Designing a control

The idea of a **control** is related to holding constant as many variables as possible. Sometimes it is appropriate to have an identical trial which leaves out the independent variable – this would be a control. A control can be compared with an experiment to ensure that changes in the dependent variable are caused by the independent variable and not some other variable.

Reliability

The concept of reliability is connected to the degree with which the results data can be trusted and the level of confidence in the results. This is often increased when an experiment is repeated several times. Repeating experiments lessens the chance of the results being **atypical** or caused by an **error** or a fluke. The data generated by repeating experiments can be averaged to simplify it to one value.

Evaluation

Evaluation is the consideration of the validity and reliability of an experiment. The validity often depends on the degree to which the experiment is fair and well controlled but it also depends on the method used, which should be clear of flaws. Reliability often depends on the number of repeats of the experiment. Exam questions on evaluation often deal with these ideas and it is common to be asked to suggest improvements to methods to correct weaknesses in validity or reliability.

Presenting information

This skill involves analysing data usually presented in the form of a table in order to present it as a line graph.

The table of data below is typical of what you could be asked to present.

Hint

Describing a control in an experiment is always tricky for candidates – make sure you learn the right words to do this.

Control

Part of an experimental procedure which can be compared to experimental situations to show that it is the independent variable that is causing any changes to the dependent variable.

Hint

In experiments which involve samples being taken, increasing the number or size of the sample can also increase reliability.

Example

The effect of temperature on the rate of respiration of a tissue sample was investigated using a respirometer to measure the volume of carbon dioxide it produced per hour.

Temperature (°C)	Rate of respiration (cm^3 of CO_2 produced per hour)
5	10
10	20
15	40
25	60
30	70

Read the hints overleaf before you try to present this data as a line graph on a separate piece of graph paper.

Data-handling skill set

Experiments generate results which are in the form of data. In many modern situations the data will be imported into databases and processed using computers but in journals and books, analysed data is generally presented in the form of graphs and charts as shown in **Figure 4.2**. In your exam there will be a long data-handling question, often with two sources of data given, worth between 5 and 9 marks. It is likely to focus on the results of an experiment which has been carried out. The question should concentrate on **selecting** and **processing** data, **concluding** and **predicting** from the data. You could be asked to present some of the data as a graph (see **Presenting information** above).

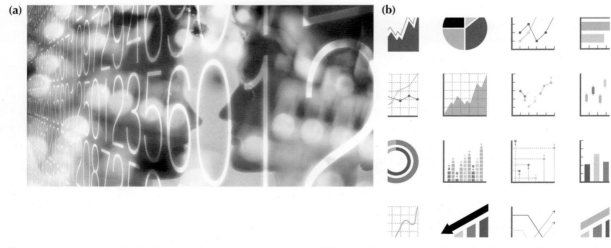

Figure 4.2 *(a) In modern biology and medicine, experimental data is often imported into databases and processed using computer technology (b) In journals and books, graphs and charts are used to display analysed data so that trends can be identified and conclusions reached*

Selecting information

Questions will often present data in the form of tables, graphs and charts. You could be asked to use values to describe the data, identify any trends which can be seen within the data or simply to select specific data.

Tables

Example

The table shows the percentage of adult male and female cigarette smokers in different age groups in the UK in 2011.

Age group	Adult smokers (% of the age group)	
	Male	Female
18–24	25·1	19·9
25–34	28·5	19·2
35–49	24·3	18·9
50–64	19·5	16·7
65+	9·5	9·0

> ### 🔍 Hint
> Because the percentage of males smoking increases between the 18–24 years group and the 25–34 group, this means that the decreasing trend cannot apply to males or to all groups.

You could be asked to identify trends shown in the data.

The percentage of females in each age group who smoke decreases as the age group gets older.

More males in every age category smoke compared to females.

Double axis graphs

Example

The graph shows the levels of blood glucose and blood insulin in an individual measured over a 150-minute period.

You could be asked to give the blood insulin concentration at 60 minutes.

Read up from 60 minutes to the insulin line identified from the key. Then go to the right y-axis scale to read the value: **400 units per litre**.

Another common question involves giving a value for one variable when the other is at a given level. You could be asked to give the insulin concentration when the blood glucose concentration is at 175 mg/100 cm^3.

Read across from 175 on the glucose scale to the glucose graph, then down at that point to the insulin graph and then across to the insulin scale: **200 units per litre**.

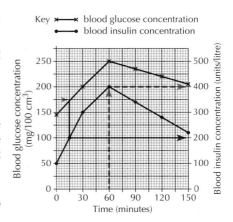

Key ✕—✕ blood glucose concentration
●—● blood insulin concentration

> ### 🔍 Hint
> Focusing all through the task is essential. It is easy to lose your way through this type of data and read up to the wrong line or go to the wrong y-axis.

> ### 🔍 Hint
> It can be useful to use highlighter pens to help you link the lines with the correct scales on a double Y axis graph.

Error bars

An error bar gives an indication of the range of the data obtained for a particular value of the independent variable. Generally, the longer the error bar, the wider the range of values obtained and so the less reliable is the average value plotted. If error bars for two data points overlap, the differences in the two data points would not be seen as significantly different.

Example

Two athletes took part in different training programmes. One undertook regular training sessions while the other undertook high-intensity interval training (HIIT) sessions. The graph shows the time taken to complete a run before and after each training programme.

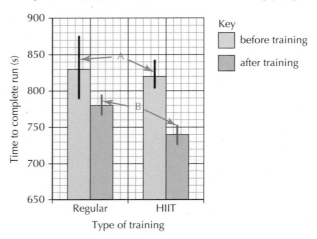

The error bars which are used with the 'before' training data show that the two athletes were not significantly different in their basic pre-training fitness because their two error bars are overlapping a lot – see **A** on the chart. However, the error bars used with the 'after' training data show that the different training regimes gave significantly different results because these error bars do not overlap each other – see **B** on the chart.

Hint

Watch for the overlap of error bars – it's crucial!

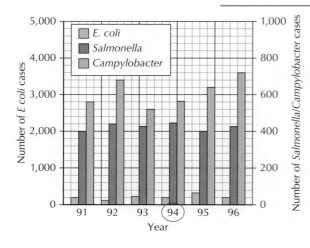

Double axis chart

Example

The chart on the left is typical and involves compound data, making it tricky to read.

The chart shows the number of cases of intestinal infections reported in a health board area between 1991 and 1996.

You could be asked to use values from the graph to compare the number of cases of each infection in 1994.

E. coli: each small division is 1/5 of 1,000 = 200 so **200 reported cases of *E. coli***.

Salmonella/Campylobacter: each small division is 1/5 of 200 = 40 so **440 cases of *Salmonella*** and **560 cases of *Campylobacter***.

Hint

It's essential to read the key properly and to use the correct scale for each bacterium – note that *E. coli* is on the left y-axis but *Salmonella* and *Campylobacter* are both on the right y-axis. Find the value of the smallest division on each scale.

Processing information

Processing usually means performing calculations. You may be asked to select the data needed for the calculation from a source such as a line graph. Again, double axis graphs are often used because the selecting skill can be an appropriate challenge.

Hint

Concentration is essential with double axis graphs and charts in order to avoid making mistakes with them.

Example

The graph shows the number of cases of measles in a developing area of Asia as well as the measles vaccination rate in the population of that area.

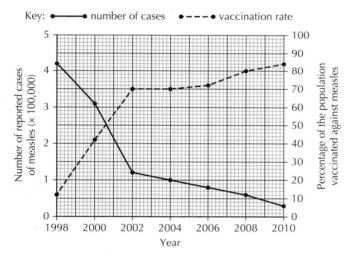

Average change with time

You could be asked to calculate the average yearly increase in the percentage of the population that are vaccinated against measles.

The data extends from 1998 to 2010 – that can be taken as 12 years.

The dotted line vaccination rate increases from 12% in 1998 to 84% in 2010 – an increase of 72%.

To find the average increase per year divide 72% by 12 = **6% average increase per year**.

Whole number ratio

You could be asked to calculate the simplest whole number ratio of the number of cases in 2002 to that in 2010.

2002 = 1·2 × 100,000 and 2010 = 0·3 × 100,000; because they are both × 100,000, this can be ignored so the ratio is 1·2 : 0·3 which can be converted to a whole number ratio by dividing each side by 0·3 = **4 : 1**.

Percentage change

You could be asked to calculate the percentage decrease in the number of reported cases across the whole period.

The number of cases started at 4·2 × 100,000 in 1998 and reduced to 0·3 × 100,000 in 2010 so the decrease in cases is 3·9 × 100,000. This is then expressed as a percentage of the starting value of 4·2 × 100,000; because they are both × 100,000, this can be ignored so the calculation is 3·9/4·2 × 100 = **92·9**. Note that this value has been rounded and you would be expected to do this.

Hint

- A calculator is highly recommended to perform calculations – make sure yours is working!
- You should be aware that rounding numbers is sometimes expected. This could be when calculations produce answers with greater precision than that of the original data.
- In general, the number of decimal points which would be sought would not exceed three places.
- In most processing questions there will be a space given for the calculation – you should use this space because you will most likely need it.
- There are often questions which require calculation of relative numbers of bases, sugars or phosphates in sections of DNA – these calculations are seen as application of knowledge. To do these successfully you need to remember the base pairing rules and the components of a single nucleotide.

Substituting into equations

Cardiac output is calculated by the following equation:

Cardiac output (cm³ per minute) = stroke volume (cm³) ×

heart rate (beats per minute)

Think about this equation as the expression: $X = Y \times Z$.

You could be asked to solve for any of the components of the equation so remember that $Y = \dfrac{X}{Z}$ and $Z = \dfrac{X}{Y}$.

Body mass index is calculated by the following equation:

$$BMI = \frac{body\ mass\ (kg)}{height\ (m)^2}$$

Think about this equation as the expression: $X = \dfrac{Y}{Z^2}$.

You could be asked to solve for any of the components of the equation so remember that $Y = X \times Z^2$ and $Z^2 = \dfrac{Y}{X}$ so $Z = \sqrt{\dfrac{Y}{X}}$. That last one is tricky!

Concluding

This skill involves looking at data and trying to summarise specific trends which the data shows to allow a statement to be made about how variables have affected each other.

Hint

If you are asked to calculate cardiac output or body mass index and also to supply the equation, then the skills being tested include application of knowledge. You are expected to know these equations.

Concluding from experimental data

If experimental data is given then summarising how the independent variable has affected the dependent variable is the basis of most conclusions. It is very important to relate any conclusion which you draw to the aim of the experiment, which is usually stated in the stem of the question. It is also important to highlight any trends in the data, even if the data is not changing.

Example

An investigation into how oxygen levels affect the rate of respiration and the uptake of potassium ions in muscle tissue was carried out and the results shown in the graph.

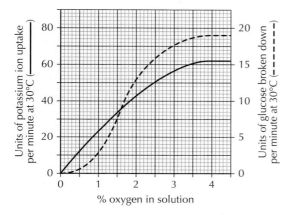

You could be asked to give a conclusion which could be drawn about the percentage of oxygen and the rate of respiration in the muscle cells.

As the percentage of oxygen in solution increased from 0 to 3·8%, the rate of respiration increased but as the percentage of oxygen increased further the rate of respiration remained the same.

🔍 Hint

- Notice that there are **two** aspects to the conclusion as seen in the data trends in this example.
- It is important to note that the rate of respiration is being measured by the units of glucose being broken down per minute.
- Conclusions about respiration rate should be linked to glucose breakdown but should be worded in terms of respiration rate because that was the stated aim. There should be no mention of the potassium uptake which is irrelevant to the conclusion sought.
- Relate your conclusion to the aim, otherwise you will receive no marks!

Concluding from observation

If the data has come from the observation of previously collected records them there will not be an aim and the conclusion will have to be worked out from trends in the data alone.

Example

The table below shows data on life expectancy in Scotland collected from census data.

	Average life expectancy of child (years)		
Year	From birth	From age 1	From age 15
1861	42	47	43
1891	46	52	46
1921	55	59	50
1951	66	68	55
1981	72	72	59
2011	78	77	63

You could be asked to give a conclusion about how child life expectancy changed between 1861 and 2011.

Life expectancy has increased for children **of all ages** investigated.

Infant mortality in the first year of life decreased between 1861 and 2011.

You could be asked to give evidence from the data to support this conclusion. This is tricky!

In 1861, 1-year-old children had an average life expectancy of a further 47 years which would mean living to 48, whereas a newborn could only expect to live until 42, showing that there must have been high mortality in the first year of life. By 2011, 1-year-old children had an average life expectancy of a further 77 years which would mean living to 78 years, exactly the same as those at birth, so mortality in the first year of life was effectively zero.

Predicting

This skill is about looking at trends in data from tables or graphs and trying to infer what the result might be for a value of the independent variable which has not been tested.

Hint

You should note that life expectancy is expressed as the average number of further years to live.

Hint

It would be essential to state that the conclusion refers to all of the ages in the data.

Interpolation from tables

Example

Yeast was grown in different lactose concentrations and the mass of alcohol it produced per 100 cm³ was measured after 12 and 36 hours as shown in the table.

Lactose concentration (%)	Ethanol concentration (g per 100 cm³)	
	12 hours	36 hours
4	1·20	1·65
8	1·55	2·80
12	2·00	4·25
16	2·80	3·25
20	2·80	6·50

You could be asked to predict the ethanol concentration after 12 hours if the yeast had been grown in 14% lactose.

Find the gap where 14% would lie – half way between 12% and 16%. Scan across to the 12 hours results between 12% and 16%.

Work out where the halfway point between the 12% and 16% data would be – in this case 2·40 g per 100 cm³.

Extrapolation from graphs

Example

The graph shows the stroke volume and heart rate for an individual as their exercise intensity increased by evenly stepped levels.

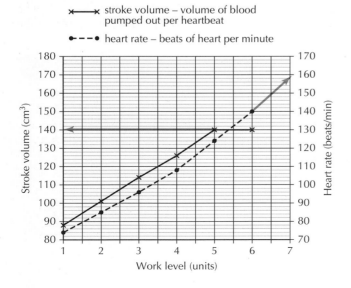

You could be asked to predict the stroke volume and heart rate if the work level was increased to 7 units.

The stroke volume is already at its maximum and has levelled out so there is no reason to think it would increase further **= 140 cm³**.

The heart rate might be expected to rise by a similar amount as when the work level increased from 5 to 6 so it would be reasonable to continue the dotted line at the same angle until the y-axis **= 158 bpm**.

Apparatus and techniques

The course specification for Higher Human Biology lists apparatus and techniques with which you should be familiar for your exam and also possibly for your assignment.

Apparatus

There is a list of apparatus in the course specification with which you are expected to be familiar for your exam. The following are items which you probably met at National 5 and which appear in various Activities throughout this book. The chapter and page where they first appear is given. You should check with your teacher if you are unsure about any of these items.

- beaker (Chapter 1, page 95)
- measuring cylinder (Chapter 1, page 31)
- dropper or pipette (Chapter 1, page 50)
- test tube or boiling tube (Chapter 1, page 31)
- funnel (Chapter 1, page 31)
- Petri dish (Chapter 1, page 48)
- water bath (Chapter 1, page 31)

The measuring devices in the table below are also mentioned by name in the course specification. You need to be familiar with what they measure and the units commonly used. The chapter and page where they first appear in this book is given.

Measuring device	What it measures (units)	Chapter	Page
Colorimeter	Concentration of a coloured dye (absorption units)	1.7	69
Pulsometer	Pulse rate (beats per minute)	2.6	155
Sphygmomanometer	Blood pressure (mm Hg)	2.6	154
Thermometer	Temperature (°C)	1.7	70
Stopwatch	Time (seconds)	1.1	19
Syringe	Volume (cm³ or millilitres)	1.6	67
Balance	Mass (grams)	4 (Q1)	275

Techniques

The techniques logo has been used throughout the book to highlight the six techniques with which you are expected to be familiar for your exam. The table summarises these techniques and gives the chapters and the pages on which they are first introduced. Also included in the table are two techniques which are mentioned in the course specification as suggested learning activities but which SQA has used to exemplify possible assignment techniques.

Technique specifically named	Chapter	Page
Using gel electrophoresis to separate macromolecules such as DNA fragments	1.2	28
Using substrate concentration or inhibitor concentration to alter reaction rates	1.6	66
Measuring metabolic rate using oxygen, carbon dioxide, temperature probes	1.6	67
Using a respirometer to measure rate of respiration	1.6	67
Measuring pulse rate and blood pressure	2.6	154
Measuring body mass index	2.8	178
Technique exemplified as an assignment technique	**Chapter**	**Page**
Using UV-sensitive yeast to measure mutation rate	1.4	48
Using the serial position effect to investigate memory	3.3	208

GO! Activity 4.1 Work individually to …

1. **Experimental skill set example**

   An investigation was carried out on the effect of inhibitor concentration on the activity of an enzyme which breaks down ethanol in liver cells.

   Six test tubes were set up each containing ethanol of the same concentration and inhibitor. One gram of liver was weighed on a balance and added to each tube as shown in the diagram.

   The tubes were placed in a water bath set at 30°C and left for one hour. The concentration of ethanol remaining in the tubes after this time was measured and expressed as a percentage of the initial ethanol concentration as shown in the table.

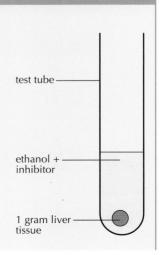

test tube

ethanol + inhibitor

1 gram liver tissue

(*continued*)

Inhibitor concentration (mM)	Final ethanol concentration (% of the initial ethanol concentration)
0.5	20
1.5	25
2.5	50
3.5	95
4.0	100
5.5	100

a) **Give two** variables not already mentioned which should be kept constant to ensure that the results of the investigation are valid. 2

b) **Describe** the controls which should be set up to show that the reactions in the tubes are caused by the liver enzyme. 1

c) On a separate piece of graph paper, **plot a line graph** of the data shown in the table. 2

d) **Use the data to describe** the relationship between the concentration of inhibitor and the activity of the enzyme. 2

e) **Evaluate** the results of the experiment by commenting on their reliability. 1

2. **Data-handling skill set example**

In a clinical trial into the effect of a new statin drug, 50 volunteers with elevated cholesterol levels were divided randomly into two groups.

The experimental Group A were given a 10 mg tablet of the new statin drug each day during a year-long trial. The control Group B received a placebo tablet each day over the same period.

The blood cholesterol level of all the individuals involved was taken on the last day of each month over the year and the averages calculated. The table shows the results. The data collected during one of the months of the trial is summarised in the bar chart.

Month of trial	Average blood cholesterol (millimoles per litre)	
	Group A	Group B
0	6.3	6.2
2	6.3	6.3
4	6.3	6.2
6	6.4	6.2
8	5.6	6.3
10	5.3	6.2
12	5.1	6.1

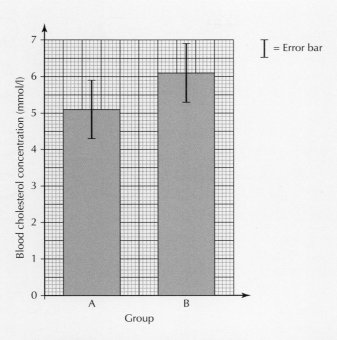

a) **Use values from the table to describe** the effect of the drug on cholesterol levels. 2

b) **Calculate** the average monthly decrease in cholesterol levels in Group A over the year. 1

c) **Calculate** the percentage decrease in cholesterol levels in Group A between 6 months and 8 months into the trial. 1

d) **Give one** conclusion which can be reached regarding the effectiveness of the drug in reducing cholesterol in the Group A volunteers. 1

e) **Predict** the average reading which would have been expected in Group A in the ninth month of the trial. 1

f) **Use evidence from the table to identify** the month of the trial from which the data in the bar chart were taken. 1

g) **Use evidence from the bar chart to explain** why it was decided that the drug might not be worth further development. 2

Learning checklist

After working on this chapter, I can:

Skills

1. *Plan and design experiments to test hypotheses, taking account of the need for control of variables and aspects of reliability.*

2. *Select information from a variety of sources, including tables and double axis graphs and charts.*

3. *Present information appropriately as a line graph.*

4. *Process information using calculations, including average change, ratios, percentage change and substituting into equations using appropriate units.*

5. *Make predictions, including interpolations and extrapolations based on evidence or by applying knowledge to enhance information supplied.*

6. *Draw valid conclusions and give explanations supported by evidence and justifications from supplied data.*

7. *Evaluate and suggest improvements to experimental methods, including making comments on validity and reliability.*

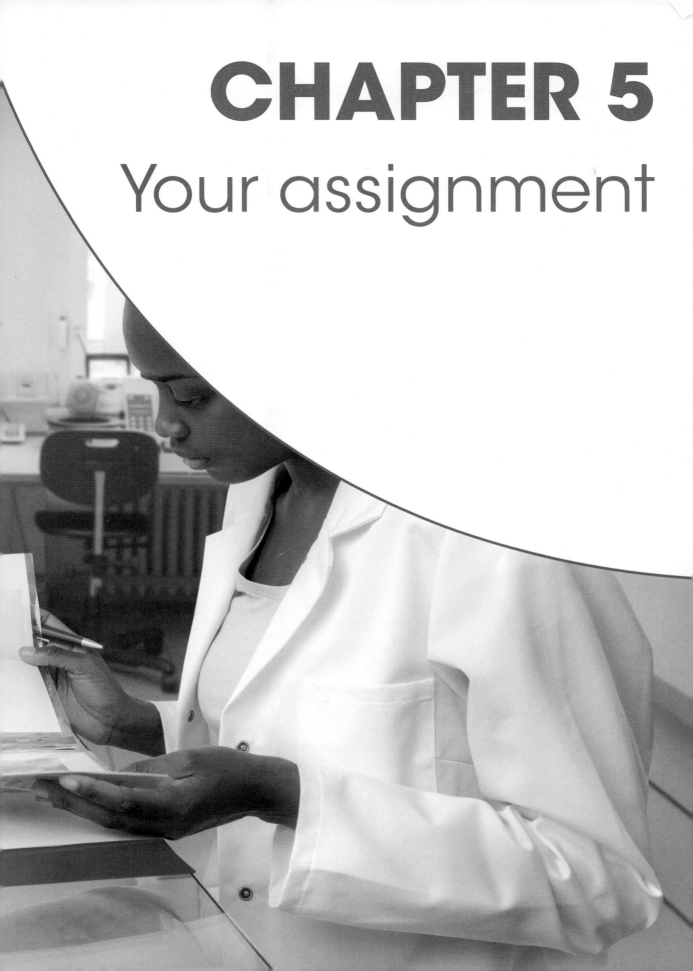

CHAPTER 5
Your assignment

Your assignment

The assignment has a total mark allocation of 20 marks. This is scaled to 30 marks by SQA to represent 20% of the overall marks for the course assessment. The remaining 80% of the marks is allocated to the two question papers that make up the examination. The assignment is an individual piece of work started at an appropriate point in the course and conducted under controlled conditions.

Candidates research and report on a topic that allows them to apply skills and knowledge in Human Biology at a level appropriate to Higher. The topic must be chosen with guidance from your teacher and must involve experimental work.

Structure of the assignment

There are two stages to the assignment. No more than 8 hours should be spent on the whole assignment.

Research stage

The research stage must involve experimental work that allows measurements to be made. Candidates must also gather data and information from the internet, books or journals. The research stage should take around 6 hours.

In the research stage, teachers must agree the chosen topic with the candidate, provide advice on the suitability of the candidate's aim and can supply instructions for the experimental procedure. Candidates must undertake research using websites, journals and/or books. A wide list of URLs and/or a wide range of books and journals may be provided by teachers. Teachers must not provide an aim, experimental data, a blank or pre-populated table for experimental results or feedback on the research. Once candidates have agreed the topic with their teacher, they must formulate an aim. After the candidate has formulated an aim, they can progress through the research stage.

Experimental work

Teachers can supply instructions for the experimental procedure but this must be only a basic list. These instructions must not include details of the number and range of values or reference to repeats. Candidates must decide on these for themselves. Candidates must carry out the experimental work individually or as part of a small group of 2–4 candidates. Within the small group, it is acceptable for candidates to share experimental data but experimental data must not be shared between groups. Where candidates identify a problem with their results and indicate that they wish to repeat the experimental work, they may do so.

Internet or literature research

This must be individual work and candidates cannot work in a group to carry out this research. Candidates may carry out research to find comparative data/information and underlying human biology outwith the direct supervision of teachers. Candidates must undertake research using only websites, journals and/or books to find secondary data/information. Candidates must find internet/literature data that they can directly compare to their experimental data and record the reference to the source. This can be data that:

- matches the sample range used

- is not an exact match for the sample range used

- is generic and illustrates a trend or pattern expected in the experimental data. Where it is not possible to find such data, candidates should aim to find information that may:

  o directly support the experimental data

  o be in contrast to the experimental data.

In circumstances where there is difficulty locating secondary data/information, teachers may provide candidates with a wide list of URLs and/or a wide range of books and/or journals. (A wide list is specified as a minimum of six.) This list must have a sufficient range of sources to allow candidates to make decisions about which data/information is relevant. Only where internet access is an issue, teachers can provide candidates with a printed copy of the full content of all URLs given in the list.

Teachers must ensure that the level of demand of the research task is the same for all candidates irrespective of the approach taken. Teachers must not provide candidates with feedback on their research.

Report stage

The report stage is conducted under a high degree of supervision and control.

Candidates are given a maximum of 2 hours to produce the report. The report is submitted to SQA for external marking. The only materials which can be used to support the report stage are:

- SQA Instructions for Candidates

- the candidate's raw experimental data

- data and information taken from the internet or literature, including a record of these source(s)

- extract(s) from the internet or literature sources to support the underlying human biology and the experimental method, if appropriate

There is no word count applied to the final report.

During the report stage candidates must not have access to:

- a previously prepared draft of a report or any part of a report
- the assignment marking instructions
- the internet
- sample calculations
- a table containing additional blank or pre-populated columns for mean and derived values

Teachers must not provide any form of feedback to a candidate on their report and, following completion, candidates must not be given an opportunity to redraft their report.

Mark allocations

The table below gives details of the mark allocation for each section of the report.

Section	Expected response	Marks
Title and structure	A clear and concise report with an informative title	1
Aim	An aim that describes clearly the purpose of the investigation	1
Underlying human biology	An account of human biology relevant to the aim of the investigation	4
Data collection and handling	A brief summary of the approach used to collect experimental data	1
	Sufficient raw data from the candidate's experiment	1
	Data, including mean values, presented in a correctly produced table	1
	Data/information relevant to the experiment obtained from an internet/literature source	1
	A citation and reference for a source of internet/literature data or information	1
Graphical presentation	An appropriate format from the options of line graph or bar graph	1
	The axes of the graph have suitable scales	1
	The axes of the graph have suitable labels and units	1
	Data points are plotted accurately with a line or clear bar tops (as appropriate)	1
Analysis	A correct comparison of the experimental data with data/information from the internet/literature source or a correctly completed calculation(s) based on the experimental data, linked to the aim	1
Conclusion	A valid conclusion that relates to the aim and is supported by all the data in the report	1
Evaluation	Evaluation of the investigation	3
TOTAL		**20**

Tips for challenging aspects of the assignment

Title and structure

You must give the title and aim for your experiment.

Aim

You must give an aim which clearly states an independent and a dependent variable which then allows you to draw a valid conclusion.

Underlying human biology

You must choose a topic directly related to the Higher Human Biology course and be sure to provide knowledge at Higher level – not National 5. The background human biology which you give must be relevant to your aim. To score full marks, you must give at least four relevant points of background human biology.

Data collection and handling

You must only provide a brief summary of the approach you used to collect experimental data – not a detailed account.

It is sensible to design and draw some table grids to take your results before you start carrying out your work.

You must have your own copy of the raw results data for the report stage.

During your report write up, you must process your data by, for example, calculating mean values and presenting the data as a line graph or bar chart as appropriate to the data itself.

Selecting information from existing sources

You must research your aim and select information from other sources such as books, journals or the internet. Your teacher will probably supply you with a minimum of six sources from which you can choose.

You must give full references for all of the sources you used.

Graphical presentation

You must transfer complete table headings, including units, to graph labels.

You must select an appropriate scale so that your graph occupies at least half of the graph paper you are using.

Analysis

The information you find has to be compared with the data you generated during your experimental work. You must describe the relationship shown by the selected or processed data clearly.

Conclusion(s)

You must give a concise conclusion related to the aim and not an account of the entire investigation. You must be careful to take account of both your experimental data and your second source of data when drawing a conclusion, even if the data is conflicting.

Note that the mark for concluding cannot be awarded if you have not stated an aim in the first place.

Evaluation

The evaluation can focus on reliability, validity and accuracy of measurements.

You should suggest an improvement that could minimise the effect of any error.

Appendix 1 Card sort template

Appendix 2 Dice and Slice board

	🎲1	🎲2	🎲3	🎲4	🎲5	🎲6
🎲1	1	2	3	4	5	6
🎲2	7	8	9	10	11	12
🎲3	13	14	15	16	17	18
🎲4	19	20	21	22	23	24
🎲5	25	26	27	28	29	30
🎲6	31	32	33	34	35	36

Player 1 scorecard

Dice numbers		Total
Overall total		

Player 2 scorecard

Dice numbers		Total
Overall total		

Appendix 3 Placemat template

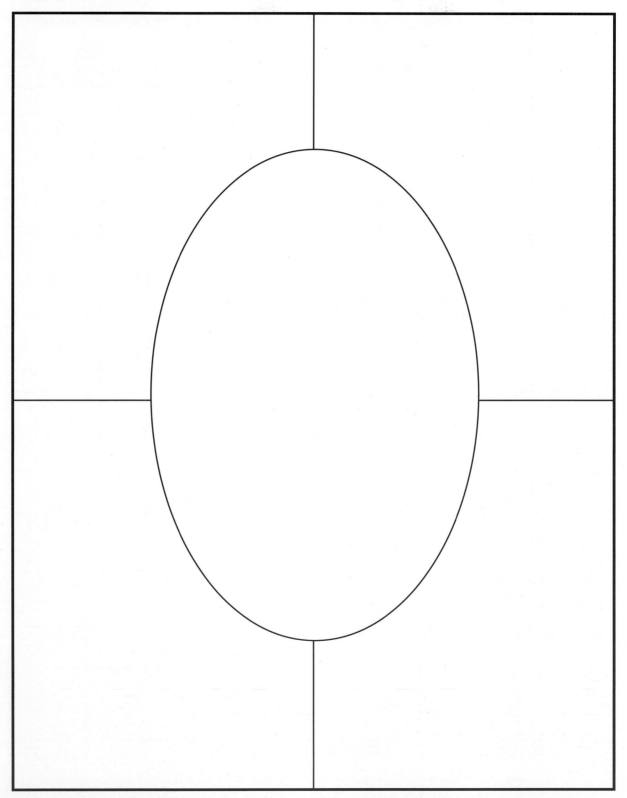

Appendix 4 Assignment planning grid

Section	Notes for my assignment	Checklist (✓)	Marks
Title and structure			1
Aim			1
Underlying human biology			4
Data collection and handling			5
Graphical presentation			4
Analysis			1
Conclusion			1
Evaluation			3
Total			20

Notes

INDEX